高等院校数据科学与大数据技术系列规划教材

大数据
离线分析

傅德谦　主编
赵向兵　张林涛　刘鸣涛　副主编

清华大学出版社
北京

内 容 简 介

本书基于开源 Hadoop 大数据生态圈的主流离线分析工具 Hive 和 Pig，通过技术讲解和案例实战相结合的方式，介绍了海量数据离线分析的技术方法。本书内容主要包括 Hive 数据库表、基于 HiveQL 的常规操作、视图、索引和 Pig 等数据处理分析和基础工具知识，Hive 函数、Pig Latin 编程、ETL 工具 Sqoop 和工作流引擎 Oozie 等相关高级技术，以及实际项目案例。

本书既可供学习大数据离线分析技术的本科和高职高专学生作为教材，也可供从事数据分析相关工作的技术人员作为参考资料。

本书封面贴有清华大学出版社防伪标签，无标签者不得销售。
版权所有，侵权必究。举报：010-62782989，beiqinquan@tup.tsinghua.edu.cn。

图书在版编目(CIP)数据

大数据离线分析/傅德谦主编. —北京：清华大学出版社，2017（2024.1重印）
（高等院校数据科学与大数据技术系列规划教材）
ISBN 978-7-302-48329-8

Ⅰ.①大… Ⅱ.①傅… Ⅲ.①数据处理 Ⅳ.①TP274

中国版本图书馆 CIP 数据核字（2017）第 215771 号

责任编辑：刘翰鹏
封面设计：傅瑞学
责任校对：袁　芳
责任印制：刘海龙

出版发行：清华大学出版社
网　　址：https://www.tup.com.cn, https://www.wqxuetang.com
地　　址：北京清华大学学研大厦 A 座
邮　　编：100084
社 总 机：010-83470000
邮　　购：010-62786544
投稿与读者服务：010-62776969, c-service@tup.tsinghua.edu.cn
质量反馈：010-62772015, zhiliang@tup.tsinghua.edu.cn
课件下载：https://www.tup.com.cn, 010-62770175-4278

印 装 者：北京建宏印刷有限公司
经　　销：全国新华书店
开　　本：185mm×260mm
印　　张：11.5
字　　数：273千字
版　　次：2017年8月第1版
印　　次：2024年1月第5次印刷
定　　价：35.00元

产品编号：076542-01

丛书编委会

(排名不分先后,按姓名汉语拼音排序)

鲍　洁	全国高等院校计算机基础教育研究会高职高专专业委员会理事长
陈文兵	南京信息工程大学数学与统计学院教授/系主任
付学良	内蒙古农业大学计算机与信息工程学院教授/副院长
高　林	全国高等学校计算机基础教育研究会副理事长
贾银山	辽宁石油化工大学计算机与通信工程学院教授/副院长
蒋　翔	广州航海学院信息与通信工程学院副教授/副院长
李辉勇	北京航空航天大学计算机学院实验师
李跃文	上海工程技术大学管理学院副教授/副院长
刘　正	苏州工业园区服务外包职业学院信息工程学院副教授/院长
罗会亮	黔南民族师范学院数学与统计学院教授/主任
马晓轩	北京建筑大学电子与信息学院院长助理
秦品乐	中北大学大数据学院副教授/副院长
盛鸿宇	北京联合大学电信实训基地高级工程师
王素贞	河北经贸大学信息技术学院院长
王业贤	东北石油大学数学与统计学院副研究员/副书记
王智萍	大唐软件技术股份有限公司智慧城市事业部副总经理
温廷新	辽宁工程技术大学工商管理学院教授/系主任
吴　斌	北京邮电大学计算机学院教授
吴　钊	湖北文理学院科研处长
肖政宏	广东技术师范学院计算机科学学院教授/副院长
熊　杰	长江大学电子信息学院副教授/系主任
徐华丽	皖西学院电子与信息工程学院副教授/网络工程实验室主任
叶　刚	北京普开数据技术有限公司 CEO
叶曲炜	哈尔滨广厦学院院长
张涵诚	东华软件股份公司大数据事业部副总经理
张晓明	北京石油化工学院信息工程学院教授/系主任

为什么要写这本书

数据时代(Data Time)的到来使大数据技术得到了学术界和产业界的重视,并获得了快速发展。随着全球数字化、移动互联网和物联网在各行各业的应用发展,使累积的数据量越来越大。诸多先行的企业、行业和国家已经证明,利用大数据技术可以更好地服务客户、发现新商业机会、扩大新市场、转换新动能。

当前正处于大数据产业发展的前期,市场需求日趋旺盛,但是人才缺口巨大,技术支撑严重不足,大数据专业知识的广泛传播非常紧迫。

本书基于教育部"2016年产学合作协同育人项目"——普开数据教学内容和课程体系改革项目,作为项目成果公开出版。北京普开数据技术有限公司在多届全国高校教师培训工作中起到了"种子"教师培养的作用,本书编者都是在培训过程中结识并展开合作的;同时在本书编写过程中,公司给予了强力支持,在此表示感谢。

读者对象

(1) 学习大数据离线分析的本科和高职高专学生。

(2) 从事数据分析相关工作的技术人员。

如何阅读本书

本书主要介绍了基于Hadoop生态圈的大数据离线处理技术。主流的大数据离线分析技术一般包括:使用HDFS存储数据,使用MapReduce做批量计算;需要数据仓库的存入Hive,从Hive进行分析和展现;涉及复杂业务场景时,使用Sqoop、Pig、Oozie等工具会更加灵活方便。

本书略过了HDFS存储数据、MapReduce批量计算的相关内容。HDFS是Hadoop提供的分布式存储框架,它可以用来存储海量数据,MapReduce是Hadoop提供的分布式计算框架,它可以用来统计和分析HDFS上的海量数据。该部分内容为Hadoop基础知识,读者如果需要深入学习,可以参考其他书籍或材料(如清华大学出版社2016年6月出版的《大数据技术基础》)。

本书内容是重点围绕Hive数据仓库展开的,Hive在Hadoop上提供了SQL接口,开发人员只需要编写简单易上手的SQL语句就可以实现创建表、删除表、加载数据、下载数据、分析数据等功能,读者可以从目录的章节名称中快速检索并学习各方面的知识。

同时,本书针对离线分析过程中的工程任务场景还提供了一些辅助工具介绍。Sqoop解决在Hadoop和关系数据库之间传递数据的问题,如果读者有这方面的基础或对其他ETL工具更熟悉,可以略过。Pig为大型数据集的处理提供了更高层次的抽象,以更灵活方便的方法实现加载数据、表达转换数据和存储最终结果,有这方面基础或暂无需求的读者可以略过书中第6、7章。Oozie实现对系统中多任务的管理,当平台中任务数量很大、需要维

护和运行时，Oozie 可以方便地完成调度监控这些任务的功能，对于仅处理简单任务场景的读者可以略过该部分内容。

偏重实践操作是本书的特色，书中所讲内容基本都配有实践操作演示。通过每部分知识的学习和相应操作环节，可以很快地掌握技术，并有很强的工程应用场景感。本书最后提供了一个综合应用案例，读者可以应用所学知识实现一个工程项目，从而有效训练工程应用开发能力。

勘误和支持

由于本书编者水平有限，书中难免会出现一些错误或者不准确的地方，恳请读者批评、指正。如果在教材使用中遇到问题，或者要学习更多相关内容，请关注微信号 lemonedu 或联系普开数据在线实验平台(lab.zkpk.org)。

编　者

2017 年 6 月

目录

绪论 ……………………………………………………………………………………… 001

第1章　走进 Hive ……………………………………………………………………… 003

 1.1　Hive 简介 ……………………………………………………………………… 003
  1.1.1　Hive 发展史 ……………………………………………………………… 003
  1.1.2　体系结构 ………………………………………………………………… 004
 1.2　Hive 的安装部署 ……………………………………………………………… 005
  1.2.1　安装配置 Hive …………………………………………………………… 005
  1.2.2　启动 Hive ………………………………………………………………… 008
 1.3　Hive 命令 ……………………………………………………………………… 009
  1.3.1　Hive 命令行选项 ………………………………………………………… 009
  1.3.2　CLI 命令行界面 ………………………………………………………… 010
  1.3.3　Hive 中 CLI 命令的快速编辑 …………………………………………… 011
  1.3.4　Hive 中的脚本 …………………………………………………………… 011
  1.3.5　dfs 命令的执行 …………………………………………………………… 013
 1.4　数据类型和文件格式 ………………………………………………………… 014
  1.4.1　基本数据类型 …………………………………………………………… 014
  1.4.2　集合数据类型 …………………………………………………………… 015
  1.4.3　文本文件数据编码 ……………………………………………………… 016
 本章小结 …………………………………………………………………………… 018
 习题 ………………………………………………………………………………… 018

第2章　HiveQL 数据定义 ……………………………………………………………… 020

 2.1　数据库的创建与查询 ………………………………………………………… 020
 2.2　数据库的修改与删除 ………………………………………………………… 021
 2.3　创建表 ………………………………………………………………………… 022
  2.3.1　管理表 …………………………………………………………………… 023
  2.3.2　外部表 …………………………………………………………………… 023
  2.3.3　查看表结构 ……………………………………………………………… 024
 2.4　修改表 ………………………………………………………………………… 025
 2.5　删除表 ………………………………………………………………………… 026

2.6 分区表 ··· 027
    2.6.1 外部分区表 ·· 028
    2.6.2 自定义表的存储格式 ·· 030
    2.6.3 增加、修改和删除分区表 ·· 031
2.7 桶表 ··· 031
本章小结 ··· 032
习题 ··· 033

第3章 HiveQL 数据操作 ··· 034

3.1 数据加载与导出 ··· 034
    3.1.1 数据加载 ·· 034
    3.1.2 数据导出 ·· 036
3.2 数据查询 ··· 037
    3.2.1 SELECT … FROM 语句 ··· 037
    3.2.2 WHERE 语句 ·· 040
    3.2.3 GROUP BY 语句与 HAVING 语句 ································ 042
    3.2.4 JOIN 语句 ·· 043
    3.2.5 ORDER BY 语句和 SORT BY 语句 ································ 046
    3.2.6 CLUSTER BY 语句 ··· 047
    3.2.7 UNION ALL 语句 ·· 048
3.3 抽样查询 ··· 048
    3.3.1 数据块抽样 ··· 049
    3.3.2 分桶表的输入裁剪 ·· 049
本章小结 ··· 051
习题 ··· 051

第4章 HiveQL 视图和索引 ·· 052

4.1 视图 ··· 052
    4.1.1 创建视图 ·· 052
    4.1.2 显示视图 ·· 053
    4.1.3 删除视图 ·· 054
4.2 索引 ··· 054
    4.2.1 创建索引 ·· 055
    4.2.2 重建索引 ·· 055
    4.2.3 显示索引 ·· 056
    4.2.4 删除索引 ·· 056
本章小结 ··· 057
习题 ··· 057

第 5 章 Hive 的函数 ································· 058

- 5.1 函数简介 ································· 058
  - 5.1.1 发现和描述函数 ················ 058
  - 5.1.2 调用函数 ······················· 059
  - 5.1.3 标准函数 ······················· 059
  - 5.1.4 聚合函数 ······················· 061
  - 5.1.5 表生成函数 ······················ 067
- 5.2 用户自定义函数 UDF ···················· 068
- 5.3 用户自定义聚合函数 UDAF ············· 072
- 5.4 用户自定义表生成函数 UDTF ··········· 074
- 5.5 UDF 的标注 ······························· 075
  - 5.5.1 定数性标注（deterministic） ···· 076
  - 5.5.2 状态性标注（stateful） ········· 076
  - 5.5.3 唯一性标注（distinctLike） ···· 076
- 本章小结 ······································· 076
- 习题 ··· 077

第 6 章 认识 Pig ································· 078

- 6.1 初识 Pig ································· 078
  - 6.1.1 Pig 是什么 ······················· 078
  - 6.1.2 Pig 的应用场景 ·················· 078
  - 6.1.3 Pig 的设计思想 ·················· 079
  - 6.1.4 Pig 的发展简史 ·················· 080
- 6.2 安装、运行 Pig ························· 080
  - 6.2.1 安装 Pig ·························· 080
  - 6.2.2 运行 Pig ·························· 081
- 本章小结 ······································· 082
- 习题 ··· 082

第 7 章 Pig 基础 ································· 084

- 7.1 命令行工具 Grunt ······················· 084
  - 7.1.1 输入 Pig Latin 脚本 ············· 084
  - 7.1.2 使用 HDFS 命令 ················ 085
  - 7.1.3 控制 Pig ·························· 087
- 7.2 Pig 数据类型 ···························· 088
  - 7.2.1 基本类型 ······················· 088
  - 7.2.2 复杂类型 ······················· 089
  - 7.2.3 NULL 值 ························ 089

7.2.4 类型转换 ………………………………………………………………… 090
本章小结 …………………………………………………………………………… 092
习题 ………………………………………………………………………………… 092

第8章　Pig Latin 编程 ……………………………………………………… 093

8.1 Pig Latin 介绍 ……………………………………………………………… 093
    8.1.1 基础知识 ……………………………………………………………… 093
    8.1.2 输入和输出 …………………………………………………………… 094
8.2 关系操作 …………………………………………………………………… 095
    8.2.1 foreach 语句 …………………………………………………………… 096
    8.2.2 filter 语句 ……………………………………………………………… 096
    8.2.3 group 语句 ……………………………………………………………… 097
    8.2.4 order 语句 ……………………………………………………………… 097
    8.2.5 distinct 语句 …………………………………………………………… 098
    8.2.6 join 语句 ……………………………………………………………… 098
    8.2.7 limit 语句 ……………………………………………………………… 098
    8.2.8 sample 语句 …………………………………………………………… 099
    8.2.9 parallel 语句 …………………………………………………………… 099
8.3 用户自定义函数 UDF ……………………………………………………… 101
    8.3.1 注册 UDF ……………………………………………………………… 102
    8.3.2 define 命令和 UDF …………………………………………………… 103
    8.3.3 调用 Java 函数 ………………………………………………………… 104
8.4 开发工具 …………………………………………………………………… 104
    8.4.1 describe ………………………………………………………………… 104
    8.4.2 explain ………………………………………………………………… 105
    8.4.3 illustrate ……………………………………………………………… 107
    8.4.4 Pig 统计信息 …………………………………………………………… 109
    8.4.5 M/R 作业状态信息 …………………………………………………… 111
    8.4.6 调试技巧 ……………………………………………………………… 112
本章小结 …………………………………………………………………………… 113
习题 ………………………………………………………………………………… 113

第9章　数据 ETL 工具 Sqoop ……………………………………………… 115

9.1 安装 Sqoop ………………………………………………………………… 115
9.2 数据导入 …………………………………………………………………… 117
    9.2.1 导入实例 ……………………………………………………………… 118
    9.2.2 导入数据的使用 ……………………………………………………… 119
    9.2.3 数据导入代码生成 …………………………………………………… 120
9.3 数据导出 …………………………………………………………………… 121

  9.3.1 导出实例 ……………………………………………………… 121
  9.3.2 导出和 SequenceFile …………………………………………… 123
本章小结 ……………………………………………………………………… 123
习题 ………………………………………………………………………… 124

第 10 章 Hadoop 工作流引擎 Oozie ……………………………………… 125

10.1 Oozie 是什么 ………………………………………………………… 125
10.2 Oozie 的安装 ………………………………………………………… 125
10.3 Oozie 的编写与运行 ………………………………………………… 131
  10.3.1 Workflow 组件 ………………………………………………… 131
  10.3.2 Coordinator 组件 ……………………………………………… 133
  10.3.3 Bundle 组件 …………………………………………………… 134
  10.3.4 作业的部署与执行 …………………………………………… 134
  10.3.5 向作业传递参数 ……………………………………………… 136
10.4 Oozie 控制台 ………………………………………………………… 136
  10.4.1 控制台界面 …………………………………………………… 136
  10.4.2 获取作业信息 ………………………………………………… 137
10.5 Oozie 的高级特性 …………………………………………………… 139
  10.5.1 自定义 Oozie Workflow ……………………………………… 139
  10.5.2 使用 Oozie JavaAPI …………………………………………… 141
本章小结 ……………………………………………………………………… 143
习题 ………………………………………………………………………… 143

第 11 章 离线计算实例 ……………………………………………………… 145

11.1 微博历史数据分析 …………………………………………………… 145
  11.1.1 数据结构 ……………………………………………………… 145
  11.1.2 需求分析 ……………………………………………………… 146
  11.1.3 需求实现 ……………………………………………………… 146
11.2 电商销售数据分析 …………………………………………………… 160
  11.2.1 数据结构 ……………………………………………………… 160
  11.2.2 需求分析 ……………………………………………………… 161
  11.2.3 需求实现 ……………………………………………………… 161
本章小结 ……………………………………………………………………… 169

参考文献 ……………………………………………………………………………… 170

绪 论

1. 大数据离线分析的知识背景

大数据技术有两个主要的方向：大数据平台相关的构建、优化、运维和监控；大数据的 ETL(Extract-Transform-Load，抽取-转换-加载)、存储、计算和分析挖掘。在大数据分析的技术中，根据数据特点和应用需求又分为离线分析和实时计算两类技术，本书是针对前者。大数据离线处理主要的特点有：数据量巨大(静态的冷数据或温数据)且保存时间长；以复杂的批量运算为主要运算类型；以应用需求为导向产生中间数据或最终结果。

大数据离线分析的场景是数据总量很大、类型复杂，但真正有价值的数据可能占比例很小，需要通过从大量不相关的、各种类型的数据中去分析、挖掘，发现新的规律和新的价值，给业务提供发展趋势、模型预测等决策支持。

大数据离线分析要解决的另一个痛点是进行快速处理和分析。由于数据量很大、逻辑复杂，一个流程需要跑几十个或上百个包，使用传统技术耗时太长，有些情况下计算结果没出来就已失效。当然，传统分析技术知识对于读者在学习本书时仍然会有很大的帮助，技术架构虽然不同，但分析方法还是相近的。

2. 大数据离线分析的应用场景

目前大数据做得比较深的行业主要还是集中在互联网、电商、金融、银行和生产制造等领域。离开业务应用场景谈大数据分析就是一个空中楼阁，每一个行业、同一个行业不同的企业，对大数据的应用需求和流程都不相同，需要懂其自身的行业、懂其自身的业务。

懂行业、深挖业务是大数据离线分析项目所必需的基础。在大数据项目的规划和落地的过程中，首先要分析业务场景有哪些？确定了场景，就基本有了调研需求细节的方向。然后，考虑这些业务场景需要哪些数据资源可以支撑？哪些数据资源是可以内部解决的，哪些数据资源是需要通过外部合作的，有哪些必要的数据资源是现在没有但是可以通过增加数据获取渠道来获取的？

大数据的应用主要分为 OLTP 和 OLAP 两种。大数据离线分析支撑 OLAP。OLTP 用于支持线上业务，需要快速响应用户请求、支持事务，对于容错性和稳定性要求非常高；OLAP 主要是离线计算，用来做数据分析，实现推荐、统计、预测、决策等业务，这也是本书讨论的主题。

对于大多数从业者来讲，一般没有太深的行业背景；当然，有行业背景的人也一般不具备大数据技术，这是实际情况。需要注意的是，大数据技术工作者要有与行业专家密切配合的思维，在组建团队和开发过程中需要时刻保持这种协作意识。

3. 大数据离线分析的工具介绍

大数据离线处理目前技术上已经成熟。Hadoop 框架是主流技术，使用 HDFS 存储数据,使用MapReduce做批量计算;需要数据仓库的存入Hive,从Hive进行分析和展现;涉及复杂业务场景时,使用Sqoop、Pig、Oozie等工具会更灵活方便。其中,MapReduce批量计算是基于Hadoop大数据离线分析的基础,HDFS是Hadoop提供的分布式存储框架,它可以用来存储海量数据;MapReduce是Hadoop提供的分布式计算框架,它可以用来统计和分析HDFS上的海量数据;Hive分析的最终执行也会转换成MapReduce的分布式批量计算过程(该部分为Hadoop基础知识,本书不做介绍)。本书重点介绍Hive相关知识和常用的大数据离线分析辅助工具。

Hive作为一种基于Hadoop的数据仓库工具,通过Hive(SQL on Hadoop)在Hadoop上提供SQL接口,开发人员只需要编写简单易上手的SQL语句就可以实现创建表、删除表、加载数据、下载数据、分析数据等功能。不同于传统数据仓库技术的,Hive会负责把SQL翻译成MapReduce,在Hadoop分布式平台上以并行化的方式运行,效率会大大提升。

Sqoop是在Hadoop和关系数据库之间传递数据的工具。通过Sqoop可以方便地将数据从关系数据库导入HDFS、Hive表,或者将数据从HDFS导出到关系数据库。就像Hive把SQL翻译成MapReduce一样,Sqoop把指定的参数翻译成MapReduce,提交到Hadoop运行,完成Hadoop与其他数据库之间的数据交换。

Pig为大型数据集的处理提供了更高层次的抽象,简化了Hadoop常见的工作任务,以更灵活方便的方法实现加载数据、表达转换数据和存储最终结果。

Oozie是实现对系统中多任务管理的工具。大数据分析、数据采集、数据交换等都是一个个的任务,这些任务中有的是定时触发,有的需要依赖其他任务来触发,当平台中有几百上千个任务需要维护和运行时,Oozie可以方便地完成调度和监控这些任务的功能。

第 1 章

走 进 Hive

本章摘要

在系统地学习了 Hadoop 之后,还需要了解 Hadoop 生态圈里面的很多组件,Hive 就是其中之一,扮演数据仓库的角色。Hive 建立在 Hadoop 集群的上层,侧重于离线分析,对存储在 Hadoop 集群上的数据提供类 SQL 的接口进行操作,实现简单的 MapReduce 统计。

本章将从 Hive 的发展史讲起,然后讲解 Hive 的安装部署,接着会对 Hive 的一些命令以及它的数据类型和文件格式进行介绍,让读者在学习本章的内容后能对 Hive 组件有一个初步的认识。

1.1 Hive 简介

Hive 是建立在 Hadoop 上的开源数据仓库基础构架,用于存储和处理海量结构化数据。作为一种可以存储、查询和分析存储在 Hadoop 中的大规模数据的机制,它提供了一系列的工具用来进行数据提取、转化、加载(ETL),定义了简单的类 SQL 查询语言(称为 HQL),允许熟悉 SQL 的用户方便地使用 Hive 查询数据;同时也允许熟悉 MapReduce 的开发者开发自定义的 Mapper 和 Reducer 来处理内建的 Mapper 和 Reducer 无法完成的复杂的分析工作。可以把 Hive 中海量结构化数据看成一张张的表,而实际上这些数据是分布式存储在 HDFS 中的。Hive 经过对语句进行解析和转换,最终生成一系列基于 Hadoop 的 Map/Reduce 任务,通过执行这些任务完成数据处理。

1.1.1 Hive 发展史

1. Hive 的诞生

Hive 诞生于 Facebook 中的日志分析需求,其设计目的是让精通 SQL 技能的分析师能够在 Facebook 存放在 HDFS 的大规模数据集上进行查询。面对海量的结构化数据,Hive 以较低的成本完成了以往需要大规模数据库才能完成的任务,并且学习门槛相对较低,应用开发灵活而高效。

2. Hive 的历史

Hive 自 2009 年 4 月 29 日发布第一个官方稳定版 0.3.0 到今天,在短短的几年时间里,一直在逐步的完善之中。从 2010 年下半年开始,Hive 成为 Apache 顶级项目。今天,Hive 已经是一个成功的 Apache 项目,很多组织把它用作一个通用的、可伸缩的数据处理平台。

1.1.2 体系结构

Hive 从外部接口中获取用户提交的 HQL 命令，然后对用户指令进行解析（需要元数据信息），实例化成一个 MapReduce 可执行计划，按照该计划生成 MapReduce 任务后交给 Hadoop 集群基于用户指定的数据进行处理，最终反馈结果给用户。

Hive 的架构如图 1-1 所示，主要由以下 4 个部分组成。

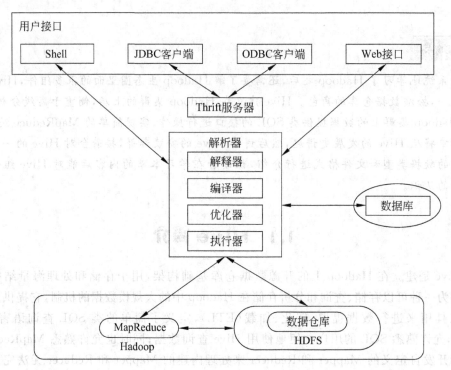

图 1-1　Hive 架构图

（1）用户接口。用户接口主要有 CLI、Client 和 Web UI。其中，最常用的是 CLI，CLI 启动时会同时启动一个 Hive 副本。Client 是 Hive 的客户端，帮助用户连接至 Hive Server。在启动 Client 模式时，需要指出 Hive Server 所在节点，并且在该节点启动 Hive Server。Web UI 提供通过浏览器访问 Hive 的方式。

（2）数据库。Hive 将元数据存储在数据库中，如 MySQL、Derby。Hive 中的元数据包括表的名字、表的列、分区及其属性、表的属性（是否为外部表等）和表的数据所在目录等。

（3）解析器。解析器包括解释器、编译器、优化器、执行器。前三个完成 HQL 查询语句从词法分析、语法分析、编译、优化以及查询计划的生成。生成的查询计划存储在 HDFS 中，并在随后由 MapReduce 调用执行。

（4）Hadoop。Hive 的数据存储在 HDFS 中，大部分的查询、计算由 MapReduce 完成。

1.2 Hive 的安装部署

1.2.1 安装配置 Hive

Hive 的安装需要在 Hadoop 已经成功安装的基础上，并且要求 Hadoop 已经正常启动（选用 Hadoop 2.7.2）。因为将 Hive 安装在 HadoopMaster 节点上，所以下面的所有操作都在 HadoopMaster 节点上进行。

Hadoop 默认的系统用户是 zkpk，密码也是 zkpk，下面所有的操作都使用 zkpk 用户，切换 zkpk 用户的命令是：

```
[zkpk@master ~]$ su - zkpk
```

1. 解压并安装 Hive

使用下面的命令解压 Hive 2.1.1 安装包。

```
[zkpk@master ~]$ cd /home/zkpk/resources/software/hadoop/apache
[zkpk@master apache]$ mv apache-hive-2.1.1-bin.tar.gz ~/
[zkpk@master apache]$ cd
[zkpk@master ~]$ tar -zxvf ~/apache-hive-2.1.1-bin.tar.gz
[zkpk@master ~]$ mv apache-hive-2.1.1-bin hive21
[zkpk@master ~]$ cd hive21
```

执行 ls -al 命令检查解压内容，会看到下面所示内容都是 Hive 包含的文件。

```
[ustb@master hive21]$ ls -al
total 108
drwxrwxr-x.  9 ustb ustb  4096 Apr 13 02:21 .
drwx------. 34 ustb ustb  4096 Apr 16 19:53 ..
drwxrwxr-x.  3 ustb ustb  4096 Apr 13 02:21 bin
drwxrwxr-x.  2 ustb ustb  4096 Apr 16 19:25 conf
drwxrwxr-x.  4 ustb ustb  4096 Apr 13 02:21 examples
drwxrwxr-x.  7 ustb ustb  4096 Apr 13 02:21 hcatalog
drwxrwxr-x.  2 ustb ustb  4096 Apr 13 02:21 jdbc
drwxrwxr-x.  4 ustb ustb 12288 Apr 16 19:26 lib
-rw-r--r--.  1 ustb ustb 29003 Nov 28 13:35 LICENSE
-rw-r--r--.  1 ustb ustb   578 Nov 29 06:09 NOTICE
-rw-r--r--.  1 ustb ustb  4122 Nov 28 13:35 README.txt
-rw-r--r--.  1 ustb ustb 18501 Nov 29 11:45 RELEASE_NOTES.txt
drwxrwxr-x.  4 ustb ustb  4096 Apr 16 19:32 scripts
[ustb@master hive21]$
```

2. 安装配置 MySQL

注意： 安装和启动 MySQL 服务需要 root 权限，因此首先要切换成 root 用户，然后用 yum 命令（需要联网）安装。

```
[zkpk@master ~]$ su root
```

输入密码：

```
zkpk
```

(1) 安装 MySQL

方法一：通过 yum 安装 MySQL。

```
[root@master zkpk]$ yum install mysql-server
[root@master zkpk]$ yum install mysql-client
[root@master zkpk]$ yum install mysql-devel
```

注意：如果命令执行过程中提示软件包已经安装了，则不需要重新安装。

方法二：通过 rpm 安装包安装 MySQL。

首先检查 MySQL 是否已经安装（注意命令中间有"|"）。

```
[root@master zkpk]$ rpm -qa | grep mysql
```

删除 Linux 上已经安装的 MySQL 相关库信息。

```
[root@master zkpk]$ rpm -e 上步查询出的结果 -nodeps
```

删除后再次通过 rpm -qa 命令检查是否卸载干净。

然后通过如下命令安装 MySQL（其中 Mysql-server-***需要写实际安装的压缩包名）。

```
[root@master zkpk]$ rpm -ivh Mysql-server-***
```

(2) 启动 MySQL 服务

安装之后，用如下方式启动服务。

```
[root@master zkpk]$ /etc/init.d/mysqld restart
```

或

```
[root@master zkpk]$ service mysqld restart
```

如果看到如下的打印输出，表示启动成功。

```
[root@master zkpk]# /etc/init.d/mysqld restart
Stopping mysqld:                                           [  OK  ]
Starting mysqld:                                           [  OK  ]
```

以 root 用户登录 MySQL（注意这里的 root 是数据库的 root 用户，不是系统的 root 用户）。默认情况下，root 用户没有密码，可以通过下面的方式登录。

```
[root@master zkpk]$ mysql -uroot
```

然后创建 Hadoop 用户并授权。

```
mysql>grant all on *.* to hadoop@'%' identified by 'hadoop';
mysql>grant all on *.* to hadoop@'localhost' identified by 'hadoop';
mysql>grant all on *.* to hadoop@'master' identified by 'hadoop';
mysql>flush privileges;
```

创建数据库。

```
mysql>create database hive21;
```

输入命令退出 MySQL。

```
mysql>quit;
```

3. 配置 Hive

进入 Hive 的配置目录,修改配置文件。

```
[zkpk@master ~]$cd /home/zkpk/hive21/conf
```

然后在该目录下创建一个新文件 hive-site.xml,命令如下:

```
[zkpk@master conf]$vim hive-site.xml
```

将下面的内容添加到 hive-site.xml 文件中。

```xml
<?xml version="1.0" encoding="UTF-8" standalone="no"?>
<?xml-stylesheet type="text/xsl" href="configuration.xsl"?><!--
<configuration>
    <property>
        <name>javax.jdo.option.ConnectionURL</name>
        <value>
        jdbc:mysql://192.168.78.128:3306/hive21?createDatabaseIfNotExist=true&amp;useUnicode=true&amp;characterEncoding=UTF-8
        </value>
        <description>
        JDBC connect string for a JDBC metastore
        </description>
    </property>
    <property>
        <name>javax.jdo.option.ConnectionDriverName</name>
        <value>com.mysql.jdbc.Driver</value>
        <description>Driver class name for a JDBC metastore
        </description>
    </property>
    <property>
        <name>javax.jdo.option.ConnectionUserName</name>
        <value>hadoop</value>
        <description>
            username to use against metastore database
        </description>
    </property>
    <property>
        <name>javax.jdo.option.ConnectionPassword</name>
        <value>hadoop</value>
        <description>
            password to use against metastore database
        </description>
    </property>
</configuration>
```

注:画下划线部分的 IP 地址是指安装 MySQL 节点的 IP。

将 MySQL 的 java connector 复制到依赖库中。

```
[zkpk@master ~]$cd /home/zkpk/resources/software/mysql
```

```
[zkpk@master mysql]$tar -zxvf mysql-connector-java-5.1.27.tar.gz
[zkpk@master mysql]$cp
~/resources/software/mysql/mysql-connector-java-5.1.27/mysql-connector-java-
5.1.27-bin.jar ~/hive21/lib/
```

使用下面的命令打开 zkpk 用户配置文件。

```
[zkpk@master ~]$vi /home/zkpk/.bash_profile
```

将下面两行配置加入环境变量。

```
export HIVE_HOME=/home/zkpk/hive21
export PATH=$PATH:$HIVE_HOME/bin
```

1.2.2 启动 Hive

从 Hive 2.1 版本开始，需要先运行 schematool 命令来执行初始化操作。初始化 Hive 在 MySQL 里的脚本 $HIVE_HOME/scripts 如下。

```
[ustb@master ~]$ schematool -initSchema -dbType mysql
which: no hbase in (/home/ustb/hive21/bin:/home/ustb/hadoop/bin:/home/ustb/hadoo
p/sbin:/usr/java/jdk1.7.0_71//bin:/home/ustb/hadoop/bin:/home/ustb/hadoop/sbin:/
usr/java/jdk1.7.0_71//bin:/usr/local/bin:/usr/bin:/bin:/usr/local/sbin:/usr/sbin
:/sbin:/home/ustb/bin:/home/ustb/bin)
SLF4J: Class path contains multiple SLF4J bindings.
SLF4J: Found binding in [jar:file:/home/ustb/hive21/lib/log4j-slf4j-impl-2.4.1.j
ar!/org/slf4j/impl/StaticLoggerBinder.class]
SLF4J: Found binding in [jar:file:/home/ustb/hadoop/share/hadoop/common/lib/slf4
j-log4j12-1.7.10.jar!/org/slf4j/impl/StaticLoggerBinder.class]
SLF4J: See http://www.slf4j.org/codes.html#multiple_bindings for an explanation.
SLF4J: Actual binding is of type [org.apache.logging.slf4j.Log4jLoggerFactory]
Metastore connection URL:        jdbc:mysql://master:3306/hive21?createDatabaseI
fNotExist=true&useUnicode=true&characterEncoding=UTF-8
Metastore Connection Driver :    com.mysql.jdbc.Driver
Metastore connection User:       hadoop
Starting metastore schema initialization to 2.1.0
Initialization script hive-schema-2.1.0.mysql.sql
Initialization script completed
schemaTool completed
```

启动 Hive 客户端。

```
[zkpk@master ~]$hive
```

出现下面的页面则表示 Hive 部署成功。

```
[ustb@master hive21]$ hive
which: no hbase in (/home/ustb/hive21/bin:/home/ustb/hadoop/bin:/home/ustb/hadoo
p/sbin:/usr/java/jdk1.7.0_71//bin:/home/ustb/hadoop/bin:/home/ustb/hadoop/sbin:/
usr/java/jdk1.7.0_71//bin:/usr/local/bin:/usr/bin:/bin:/usr/local/sbin:/usr/sbin
:/sbin:/home/ustb/bin:/home/ustb/bin)
SLF4J: Class path contains multiple SLF4J bindings.
SLF4J: Found binding in [jar:file:/home/ustb/hive21/lib/log4j-slf4j-impl-2.4.1.j
ar!/org/slf4j/impl/StaticLoggerBinder.class]
SLF4J: Found binding in [jar:file:/home/ustb/hadoop/share/hadoop/common/lib/slf4
```

```
j-log4j12-1.7.10.jar!/org/slf4j/impl/StaticLoggerBinder.class]
SLF4J: See http://www.slf4j.org/codes.html#multiple_bindings for an explanation.
SLF4J: Actual binding is of type [org.apache.logging.slf4j.Log4jLoggerFactory]
Logging initialized using configuration in jar:file:/home/ustb/hive21/lib/hive-c
ommon-2.1.1.jar!/hive-log4j2.properties Async: true
Hive-on-MR is deprecated in Hive 2 and may not be available in the future versio
ns. Consider using a different execution engine (i.e. tez, spark) or using Hive
1.X releases.
hive>
```

1.3 Hive 命令

$HIVE_HOME/bin/hive 这个 shell 命令（以后省略为 hive）是通向命令行界面，也就是 CLI 和 Hive 服务的通道。

如果用户已经将 $HIVE_HOME/bin 加入环境变量 PATH 中，则用户只需要在 shell 提示符中输入 hive，就可以使用户的 shell 环境（bash 环境）找到这个命令。

1.3.1 Hive 命令行选项

用户执行下面的命令，可以查看到 Hive 命令的一个简单说明的选项列表，下面是 Hive v2.1.* 系列版本的输出。

```
[zkpk@master ~]$hive -help

Usage ./hive <parameters> --service serviceName <service parameters>
Service List: beeline cleardanglingscratchdir cli hbaseimport hbaseschematool he
lp hiveburninclient hiveserver2 hplsql hwi jar lineage llapdump llap llapstatus
metastore metatool orcfiledump rcfilecat schemaTool version
Parameters parsed:
  --auxpath : Auxillary jars
  --config : Hive configuration directory
  --service : Starts specific service/component. cli is default
Parameters used:
  HADOOP_HOME or HADOOP_PREFIX : Hadoop install directory
  HIVE_OPT : Hive options
For help on a particular service:
  ./hive --service serviceName --help
Debug help:  ./hive --debug --help
```

要留意的是 Service List 后面的内容。这里提供了几个服务，其中包括常用的 CLI。可以通过--service name 服务名称来启动某个服务。

表 1-1 描述了最有用的 Hive 服务。

表 1-1 最有用的 Hive 服务

选项	名称	描述
cli	命令行界面	用户定义表，执行查询等。如果没有指定其他的服务，这个是默认的服务
hiveserver	Hive Server	监听来自其他进程的 Thrift 连接的一个守护进程
hiveserver2	Hive Server 升级	是 hiveserver 的升级版，解决了 daemon 不稳定，并发请求，session 管理等问题
hwi	Hive Web 界面	是一个可以执行查询语句和其他命令的简单的 Web 界面，这样可以不用登录到集群中的某台机器上使用 CLI 来进行查询

续表

选项	名称	描述
jar	与 Hadoop jar 等价的 Hive 的接口	hadoop jar 命令的一个扩展，这样可以执行需要 Hive 环境的应用
metastore	让 metastore 作为单独的（远程）进程执行	启动一个扩展的 hive 元数据服务，可以供多客户端使用
rcfilecat	RCFile 格式	一个可以打印出 RCFile 格式文件内容的工具

--auxpath 选项：允许用户指定一个以冒号分隔的"附属的"Java 包(jar)，这些文件中包含有用户可能需要的自定义扩展等。

--config 文件目录：允许用户覆盖 $HIVE_HOME/conf 中默认的属性配置，而指向一个新的配置文件目录。

1.3.2 CLI 命令行界面

CLI 就是命令行界面，是和 Hive 交互的最常用的方式。用户可以使用 CLI 创建表、检查模式和查询表等。

1. CLI 选项

下面的命令显示了 CLI 所提供的选项列表。下面显示的是 Hive v0.13.* 版本的输出。

```
[zkpk@master ~]$hive --help --service cli
-d,--define <key=value>          Variable subsitution to apply to hive
                                 commands. e.g. -d A=B or --define A=B
   --database <databasename>     Specify the database to use
-e <quoted-query-string>         SQL from command line
-f <filename>                    SQL from files
-H,--help                        Print help information
   --hiveconf <property=value>   Use value for given property
   --hivevar <key=value>         Variable subsitution to apply to hive
                                 commands. e.g. --hivevar A=B
-i <filename>                    Initialization SQL file
-S,--silent                      Silent mode in interactive shell
-v,--verbose                     Verbose mode (echo executed SQL to the
                                 console)
```

hive --help 命令也可以简写为 hive -h，下面详细介绍这些选项。

2. 变量和属性

--define key=value 和--hivevar key=value 是等价的。它们都可以让用户在命令行定义用户自定义变量，以便于在 Hive 脚本中引用，以此满足不同情况的执行，但是此功能只有 Hive v0.8.0 版本和之后的版本才支持。

当用户使用此功能时，Hive 会将这些键-值对放在 hivevar 命令空间，这样可以和其他的内置命名空间(hiveconf、system 和 env)区分开。

注意：变量或者属性是在不同的上下文中使用的术语，但是在大多数情况下它们的功能是一样的。

表 1-2 描述了 4 种命名空间选项。

表 1-2　Hive 中变量和属性命名空间

命名空间	使用权限	描述
hivevar	可读/可写	用户自定义变量
hiveconf	可读/可写	Hive 相关的配置属性，如 hive-site.xml 下面的配置变量值
system	可读/可写	Java 定义的配置属性，如 JVM 的运行环境
env	只可读	Shell 环境（bash）定义的环境变量，如 HADOOP_HOME 等

1.3.3　Hive 中 CLI 命令的快速编辑

1．查看操作历史命令

用户可以使用上下箭头来滚动查看之前执行过的命令。Hive 会将最近执行过的 10000 行命令记录到文件 $HOME/.hivehistory 中。每一行之前的输入都是单独显示的，CLI 不会把多行命令和查询作为一个单独的历史条目。

如果用户想再次执行之前的某条命令，只需要将光标滚动到那条记录处按 Enter 键就可以了。如果用户需要修改这行记录之后再执行，就需要使用左右方向键将光标移动到需要修改的地方然后重新编辑修改就可以了。修改完成后，用户直接按 Enter 键就可以提交这条命令，无须切换到命令尾。

注意：大多数的导航按键使用的 Ctrl+字母的命令和 bash shell 中是一样的，例如，Ctrl+A 代表光标移到行首，Ctrl+B 代表光标移动到行尾。然而，类似的"元操作"Option 或者 Esc 键就不起作用了，例如，Option+F 一次向前移动一个单词这样的命令。Delete 键会删除光标左边的字符，而 Forward Delete 键不会删除掉光标当前所在的字符。

2．CLI 支持命令自动补全

如果用户在输入的过程中按 Tab 制表键，CLI 就会自动补全可能的关键字或者函数名。例如，如果用户输入 INS，然后按 Tab 键，CLI 将自动补全这个词为 INSERT；如果用户在提示符后面直接按 Tab 键，那么用户会看到如下信息：

```
hive>
Display all 444 possibilities? (y or n)
```

注意：当向 CLI 中输入语句时，如果某些行以 Tab 键开头，就会产生一个令人困惑的错误。用户这时会看到一个"是否显示所有可能的情况"的提示，此时输入流后面的字符会被认为是对这个提示的回复，也会导致命令执行失败。

1.3.4　Hive 中的脚本

Hive 可以将结构化的数据文件 hive 映射为一张数据库表，并提供几乎完整的 SQL 查询功能。Hive 可以将 SQL 语句转换为 MapReduce 任务来执行。Hive 目前还不支持像 MySQL 那样的 SQL 脚本，所以如果遇到需要批量处理 HQL 命令就相对麻烦，要使用相对比较笨的 shell 脚本执行批量 HQL 命令。它的原理很简单，在 shell 脚本中用 echo 命令将 HQL 命令以字符串的形式导入 Hive 客户端执行，还可以用重定向的方式将执行结果保存到本地文件。

1．执行 shell 命令

用户在执行简单的 bash shell 命令时，不需要退出 hive CLI 就可以执行，只需要在命令前加上叹号(!)并且以分号(;)结尾。

【例 1-1】

```
hive> ! /bin/echo "what up dog";
"what up dog"
hive> ! pwd;
hive> /home/zkpk/hadoop-2.5.1/share/hadoop/mapreduce
```

注意：Hive CLI 中不能使用需要用户进行输入的交互式命令，而且不支持 shell 的"管道"功能和文件名的自动补全功能。例如，! ls *.hql 这个命令表示的是查找文件名为 *.hql 的文件，而不是表示显示以 .hql 结尾的所有文件。

2. Hive 脚本中的注释

对于 Hive 2.1.1 版本，用户可以使用以--开头的字符串来表示注释。

【例 1-2】

```
hive> --this is a comment!
hive> select * from students.student_info;
OK
1       xiaoming        14
2       xiaohong        15
3       lining  11
4       fdsljk  12
Time taken: 0.101 seconds, Fetched: 4 row(s)
hive>
```

注意：CLI 是不会解析这些注释行的。如果用户在 CLI 中粘贴这些注释语句，将会有错误信息。它们只能放在脚本中通过 hive -f script_name 的方式执行。

3. 显示字段名称

当 Hive 中的表比较多时，很容易忘记某些字段的含义。可以让 CLI 打印出字段名称（这个功能默认是关闭的）。通过设置 hiveconf 配置项 hive.cli.print.header 为 true 来开启这项功能。

【例 1-3】

在关闭的情况下，查询语句显示如下：

```
hive> select * from mytable;
OK
name1   10
name2   20
name3   30
name4   40
name5   50
name1   10
name2   20
name3   30
name4   40
name5   50
Time taken: 1.656 seconds, Fetched: 10 row(s
```

在执行命令 set hive.cli.print.header=true 后，再次查询显示如下：

```
hive> select * from mytable;
OK
mytable.name    mytable.age
name1   10
name2   20
name3   30
name4   40
```

```
name5    50
name1    10
name2    20
name3    30
name4    40
name5    50
Time taken: 0.38 seconds, Fetched: 10 row(s)
```

如果用户希望总是看到字段名称，那么只需要将 hive.cli.print.header=true 添加到 $HOME/.hiverc 文件中。

1.3.5 dfs 命令的执行

用户可以在 Hive CLI 中执行 Hadoop 的 dfs 命令，只需要将 hadoop 命令中的关键字 hadoop 去掉，然后分号结尾，使用方法如下：

```
hive> dfs -ls /;
Found 13 items
drwxr-xr-x   - zkpk supergroup          0 2015-11-05 11:46 /0902
drwxr-xr-x   - zkpk supergroup          0 2015-11-03 10:34 /0919output-06
drwxr-xr-x   - zkpk supergroup          0 2015-10-19 15:56 /blog
drwxr-xr-x   - zkpk supergroup          0 2015-10-20 22:05 /cars
drwxr-xr-x   - zkpk supergroup          0 2015-10-20 22:17 /farm
drwxr-xr-x   - zkpk supergroup          0 2015-10-26 13:15 /hbase
drwxr-xr-x   - zkpk supergroup          0 2015-11-03 10:32 /input
drwxr-xr-x   - zkpk supergroup          0 2015-10-20 22:01 /sogou
-rw-r--r--   1 zkpk supergroup  573670020 2015-10-21 15:04 /sogou500w
drwxr-xr-x   - zkpk supergroup          0 2015-10-20 17:49 /test0925
drwxr-xr-x   - zkpk supergroup          0 2015-10-20 17:47 /testpar
drwx------   - zkpk supergroup          0 2015-10-19 22:05 /tmp
drwxr-xr-x   - zkpk supergroup          0 2016-03-04 16:32 /user
```

这种在 Hive 中使用 hadoop 命令的方式实际上比与其等价的在 bash shell 中执行的 hadoop dfs 命令要更高效。因为在 Hadoop 中每次使用这些命令都会启动一个新的 JVM 实例，而 Hive 会在同一个进程中执行这些命令。

用户可以通过如下的命令查看 dfs 所提供的所有功能选项列表。

```
hive> dfs -help;
Usage: hadoop fs [generic options]
        [-appendToFile <localsrc> ... <dst>]
        [-cat [-ignoreCrc] <src> ...]
        [-checksum <src> ...]
        [-chgrp [-R] GROUP PATH...]
        [-chmod [-R] <MODE[,MODE]... | OCTALMODE> PATH...]
        [-chown [-R] [OWNER][:[GROUP]] PATH...]
        [-copyFromLocal [-f] [-p] <localsrc> ... <dst>]
        [-copyToLocal [-p] [-ignoreCrc] [-crc] <src> ... <localdst>]
        [-count [-q] <path> ...]
        [-cp [-f] [-p | -p[topax]] <src> ... <dst>]
        [-createSnapshot <snapshotDir> [<snapshotName>]]
        [-deleteSnapshot <snapshotDir> <snapshotName>]
        [-df [-h] [<path> ...]]
        [-du [-s] [-h] <path> ...]
        [-expunge]
        [-get [-p] [-ignoreCrc] [-crc] <src> ... <localdst>]
        [-getfacl [-R] <path>]
        [-getfattr [-R] {-n name | -d} [-e en] <path>]
```

至于对这些命令的学习，可以查看自己所使用的 Hadoop 发行版的文档来进行学习。

1.4 数据类型和文件格式

Hive 支持关系型数据库中的大多数基本数据类型,同时也支持关系型数据库中很少出现的 3 种集合数据类型,下面简单地描述一下为什么要这样做。

其中一个需要考虑的因素就是这些数据类型是如何在文本文件中进行表示的,同时还要考虑文本存储中为了解决各种性能问题以及其他问题有哪些替代的方案。和大多数的数据库相比较,Hive 具有一个独特的功能——对于数据在文件中的编码方式具有一定程度的灵活性。大多数的数据库对数据具有完全的控制,这种控制既包括对数据存储到磁盘的过程的控制,也包括对数据生命周期的控制。Hive 将这些方面的控制权转交给用户,以便用户更加容易地使用各种各样的工具来管理和处理数据。

1.4.1 基本数据类型

Hive 支持多种不同长度的整型和浮点型数据类型、布尔类型、无长度限制的字符串类型、时间戳数据类型、二进制数组数据类型。

表 1-3 列举出了 Hive 所支持的基本数据类型。

表 1-3 Hive 所支持的基本数据类型

数据类型	长度	例子
TINYINT	1byte 有符号整数	20
SMALLINT	2byte 有符号整数	20
INT	4byte 有符号整数	20
BIGINT	8byte 有符号整数	20
BOOLEAN	布尔类型,true 或者 false	true
FLOAT	单精度浮点数	3.14159
DOUBLE	双精度浮点数	3.14159
STRING	字符序列,可以指定字符集,可以使用单引号或者双引号	'this is a table',"this is a table"
TIMESTAMP	整数,浮点数或者字符串	1327882394（UNIX 新纪元秒）、1327882394.123456789（UNIX 新纪元秒并跟随有纳秒数）和 '2016-03-07 12:34:56.123456789'
BINARY	字节数组	

和 SQL 语言一样,以上这些都是保留字。需要注意的是,这些所有的数据类型都是对 Java 中的接口的实现,因此这些数据类型的具体行为细节和 Java 中对应的数据类型是完全一致的。例如,STRING 类型实现的是 Java 中的 string,FLOAT 实现的是 Java 中的 float 等。

Hive 中不支持限制最大长度的"字符数组"(也就是很多字符串)类型的数据。在 Hive 中,不一定拥有数据文件,但必须能够支持使用不同的文件格式,Hive 根据不同字段间的分隔符来对其进行判断。同时,Hadoop 和 Hive 强调优化磁盘的读和写的性能,而限制列的值的长度相对来说并不重要。

新增加的数据类型 TIMESTAMP 的值可以是整数,就是距离 UNIX 新纪元时间(1970 年 1 月 1 日,午夜 12 点)的秒数;也可以是浮点数,即距离 UNIX 新纪元时间的秒数,精确到纳秒(小数点后保留 9 位数);还可以是字符串,即 JDBC 所约定的时间字符串格式,格式为 YYYY-MM-DD hh:mm:ss.ffffffff。TIMESTAMP 表示的是 UTC 时间,Hive 本身还提供了不同时区间相互转化的内置函数,在后面的章节会接触到(to_utc_timestamp 函数和 from_utc_timestamp 函数)。

> **趣味阅读**
>
> **什么是 UTC 时间?**
>
> 协调世界时(UTC)又称世界统一时间、世界标准时间、国际协调时间,简称 UTC。它从英文 Coordinated Universal Time、法文 Temps Universel Cordonné 而来。
>
> 协调世界时是以原子时秒长为基础,在时刻上尽量接近于世界时的一种时间计量系统。

数据类型 BINARY 和很多关系型数据库中的 VARBINARY 数据类型相似,但它和 BLOB 数据类型并不相同。因为 BINARY 的列是存储在记录中的,而 BLOB 不同。BINARY 可以在记录中包含任意字节,可防止 Hive 将其作为数字、字符串等进行解析。用户需要注意的是,如果用户的目标是省略掉每行记录的尾部,那么就没有必要使用 BINARY 数据类型。如果一个表的表结构指定的是 3 行,而实际数据文件每行记录包含有 5 个字段,那么在 Hive 中最后的两列数据将会被省略。

注:BLOB(Binary Large Object)是二进制大对象,是一个可以存储二进制文件的容器。

1.4.2 集合数据类型

Hive 中的列支持用户使用 STRUCT、MAP 和 ARRAY 等集合数据类型。

表 1-4 描述的是 3 种集合数据类型。需要注意的是,表中的语法实例实际上调用的是内置函数。

表 1-4 集合数据类型

数据类型	描述	字面语法示例
STRUCT	和 C 语言中的 struct 或者"对象"类似,都可以通过"点"符号访问元素内容。例如,如果某个列的数据类型是 STRUCT{field1 STRING,field2 STRING},那么第 1 个元素可以通过字段名 field1 来引用	struct('John','Doe')
MAP	MAP 是一组键-值对元组集合,使用数组表示法(例如 ['key'])可以访问元素。例如,如果某个列的数据类型是 MAP,其键→值对是'first'→'John'和'last'→'Doe',那么可以通过字段名['last']获取最后 1 个元素	map('first','JOIN','last','Doe')

续表

数据类型	描 述	字面语法示例
ARRAY	数组是一组具有相同类型和名称的变量的集合。这些变量称为数组的元素,每个数组元素都有一个编号,编号从零开始。例如,数组值为['John','Doe'],那么第 2 个元素可以通过数组名[1]进行引用	Array('John','Doe')

和基本数据类型一样,这些类型的名称也是保留字。

大多数的关系型数据库并不支持这些集合数据类型,是因为使用它们会趋向于破坏标准格式。例如,在传统的数据模型中,structs 可能需要多个不同的表拼装而成,表间需要适当地使用外键来进行连接。然而,Hive 中没有键的概念,用户可以根据需要对表建立索引。这个部分会在第 4 章进行学习。

【例 1-4】 为了更好地学习这些数据类型,创建一张虚构的人力资源应用程序中的员工表,演示如何使用这些数据类型的表结构声明语句。

```
CREATE TABLE employees(
    name                STRING,
    salary              FLOAT,
    subordinates        ARRAY<STRING>,
    deductions          MAP<STRING,FLOAT>,
    address             STRUCT<street:STRING,city:STRING,state:STRING,zip:INT>);
```

在这张表中,name 是一个简单的字符串;对于大多数雇员来说,salary(薪水)用 FLOAT 浮点数类型来表示;subordinates(下属员工)列表是一个字符串数组。在这张表中,可以选择 name 作为"主键",因此 subordinates 中的每一个元素都将会引用这张表中的另一条记录。对于那些没有下属的雇员,这个字段对应的值就是一个空的数组。在传统的模型中,将会以另外一种方式来表示这种关系,也就是雇员与雇员的经理这种对应的关系。

字段 deductions 是一个由键-值对构成的 MAP。其记录了每一次的扣除额,这些钱将会在发薪水时从员工的工资中扣除。MAP 中的键是扣除金额项目的名称(如,"国家税收"),而且键名既可以是一个百分比值,也可以完全就是一个数值。

字段 address 是雇员的家庭住址,使用的是 STRUCT 数据类型存储,其中的每个域都被作了命名,并且具有一个特定的类型。需要注意的是,如何使用 Java 语法惯例来表示集合数据类型。例如,MAP<STRING,FLOAT>表示 MAP 中的每个键都是 STRING 数据类型,而每个值都是 FLOAT 数据类型。对于 ARRAY<STRING>,其中的每个条目都是 STRING 类型。STRUCT 可以混合多种不同的数据类型,但是在 STRUCT 中一旦声明好结构,那么它的位置就不能改变。

1.4.3 文本文件数据编码

在学习文本文件数据编码之前,首先要弄清楚文件格式有哪些。就目前来说,用户很熟悉以逗号或者制表符分隔的文本文件,也就是逗号分隔值(CSV)或者制表符分隔值(TSV)。Hive 支持这些文件格式,然而,这两种文件格式有一个共同的缺点,就是用户需要对文本文件中那些不需要作为分隔符处理的逗号或制表符格外小心。因此,Hive 默认使用了几个控制字符,这些字符很少出现在字段值中。

表 1-5 列举了几种分隔符。

表 1-5 Hive 中默认的记录和字段分隔符

分隔符	描 述
\n	对于文本文件来说,每行都是一条记录,因此换行符可以分隔记录
^A(Ctrl+A)	用于分隔字段(列)。在 CREATE TABLE 语句中可以使用八进制编码\001 表示
^B	用于分隔 ARRAY 或者 STRUCT 中的元素,使用在 MAP 中键-值对之间。在 CREATE TABLE 语句中可以使用八进制编码\002 表示
^C	用于 MAP 中键和值之间的分隔。在 CREATE TABLE 语句中可以使用八进制\003 表示

例 1-4 创建的表 employees 结构没有分隔符,接下来在此建表语句中明确指定分隔符如下:

```
CREATE TABLE employees(
    name              STRING,
    salary            FLOAT,
    subordinates      ARRAY<STRING>,
    deductions        MAP<STRING,FLOAT>,
    address STRUCT<street:STRING,city:STRING,state:STRING,zip:INT>)
    ROW FORMAT DELIMITED
    FIELDS TERMINATED BY '\001'
    COLLECTION ITEMS TERMINATED BY '\002'
    MAP KEYS TERMINATED BY '\003'
    LINES TERMINATED BY '\n'
    STORED AS TEXTFILE;
```

ROW FORMAT DELIMITED 这组关键字是用来设置创建的表在加载数据时支持的列分隔符,必须要写在其他子句(除了 STORED AS...子句)之前。字符\001 是^A 的八进制数。ROW FORMAT DELIMITED FIELDS TERMINATED BY '\001'这个句子表明 Hive 将使用^A 字符作为列分隔符。字符\002 是^B 的八进制数。ROW FORMAT DELIMITED COLLECTION ITEMS TERMINATED BY '\002'这个句子表明 Hive 将使用^B 作为集合元素间的分隔符。同样,字符\003 是^C 的八进制数。ROW FORMAT DELIMITED MAP KEYS TERMINATED BY '\003',这个句子表明 Hive 将使用^C 作为 map 的键和值之间的分隔符。子句 LINES TERMINATED BY '...'和 STORED AS...不需要 ROW FORMAT DELIMITED 关键字。

然而,Hive 到目前为止,对于 LINES TERMINATED BY...仅支持字符'\n',也就是说行与行之间只能用'\n'作为分隔符,因此这个子句现在使用起来还是有限制的。当然,用户可以重新指定列分隔符及集合元素间分隔符,而 map 中键-值间分隔符仍然使用默认的文本文件格式,因此 STORED AS TEXTFILE 子句很少被使用到。本书中大多数情况下,使用的都是默认的 TEXTFILE 文件格式。

虽然用户可以明确指定这些子句,但是大多数情况下,大多子句还是使用默认的分隔符,只需要明确指定哪些是需要替换的分隔符就可以。如下表结构声明定义中,表数据按照逗号进行分隔。

```
CREATE TABLE some_data(
    first FLOAT,
    second FLOAT,
    third  FLOAT)
ROW FORMAT DELIMITED
FIELDS TERMINATED BY ',';
```

用户还可以使用'\t'(制表键)作为字段分隔符。

趣味阅读

读 时 模 式

　　数据库对于存储都具有完全的控制力。数据库就是"守门人"。传统数据库是写时模式，就是数据在写入数据库时对模式进行检查。

　　Hive对底层存储并没有这样的控制，Hive对于要查询的数据，有很多方式进行创建、修改，甚至损坏。因此，Hive不会在数据加载时进行验证，而是在查询时进行，也就是读时模式。

　　如果模式和文件内容并不匹配将会怎样呢？Hive可以读取这些数据。如果每行记录中的字段个数少于对应的模式中定义的字段个数，那么用户将会看到查询结果中有很多NULL值；如果某些字段是数值型的，但是Hive在读取时发现存在非数值型的字符值，那么那些字段将会返回NULL值。除此之外，Hive都极力尝试尽可能地将各种错误恢复过来。

本章小结

本章主要是为了让读者简单地了解一下Hive。

(1) 了解Hive的由来和组成结构，并且深入了解Hive内部是如何工作的。

(2) 独立完成Hive的安装部署，并且可以启动它。

(3) 初步了解Hive的命令，为后续更加深入学习Hive打下基础。

(4) 对Hive的数据类型有简单的了解，如果有能力最好是记住。另外，学习文本文件格式对之后的学习也有一定的帮助。

习　题

1. 选择题

(1) 以下不属于Hive体系结构的是(　　)。

　　A. 用户接口　　　B. 元数据存储　　　C. Zookeeper　　　D. 解析器

(2) 启动Hive的命令是(　　)。

A. sbin/start-all.sh　　　　　B. hive

C. sbin/stop-all.sh　　　　　 D. 以上命令都可以

（3）以下选项中,显示列表名称的设置是(　　)。

A. hive.cli.print.header=true

B. hive.cli.print.row.to.vertical=true

C. hive.cli.print.row.to.vertical.num=1

D. 以上都不正确

（4）下列集合数据类型中,不属于 Hive 的是(　　)。

A. STRING　　　B. STRUCT　　　C. ARRAY　　　D. MAP

（5）Hive 的变量和属性命令空间中,使用权限为只可读的是(　　)。

A. hivevar　　　B. hiveconf　　　C. system　　　D. env

2. 问答题

（1）Hive 是什么？它主要的职责是什么？

（2）简述 Hive 的安装步骤。

（3）Hive 服务有哪些？简单介绍一个或两个。

（4）简单描述一下 Hive 的架构,并画出 Hive 的架构图。

第 2 章 HiveQL 数据定义

本章摘要

HiveQL 是 Hive 查询语言,大部分的 HiveQL 与标准 SQL 语言相似,甚至有些语句与 MySQL 语句的用法是一致的,但是它不完全遵守任一种 ANSI SQL 标准的修订版,是因为 Hive 不支持行级插入操作、更新操作和删除操作,也不支持事务;Hive 增加了在 Hadoop 环境下更高性能的扩展,以及一些个性化的特征,甚至还增加了一些外部的程序。

本章和随后几章都采用实例的形式来进行学习。首先,学习数据库的创建和查看;其次,学习数据库的修改与删除;最后,学习 Hive 中的表(主要包括管理表、外部表、分区表以及桶表等)的创建、查看、修改和删除。

2.1 数据库的创建与查询

Hive 中数据库的概念本质上仅是一个目录或者命名空间。对于具有很多组和用户的大集群来说,这是非常有用的,因为这样可以避免表命名冲突。Hive 通常会使用数据库来将生产表组织成逻辑组。

如果用户没有显式地指定数据库,那么将会使用默认的数据库 default。

1. 创建数据库

在 Hive 中创建数据库的方法类似于 MySQL 中的创建方法。例如,在 Hive 中创建数据库 finacials。

```
hive>CREATE DATABASE financials;
```

这种创建数据库的方式适用于该数据库不存在的情况。如果此数据库已经存在,在创建时会抛出错误信息,可以使用如下语句解决这个问题。

```
hive>CREATE DATABASE IF NOT EXISTS financials;
```

IF NOT EXISTS 这个子句对于那些在继续执行之前需要实时创建数据库的情况来说是非常有用的。但是在通常情况下,用户还是希望在同名数据库已经存在的情况下能够抛出警告信息。

2. 查询数据库

在 Hive 中查询数据库的基本语句。

```
Hive>SHOW DATABASES;
```

显示如下结果。

```
hive> SHOW DATABASES;
OK
default
financials
```

在数据库非常多的场景中,上面这种查询语句就不合理了,需要使用正则表达式来筛选出需要的数据库名。例如,如何列出所有以字母 h 开头、以其他字符结尾(即 .* 部分含义)的数据库。

首先,创建以字母 h 开头的数据库。

```
hive>CREATE DATABASE human_resources;
```

查询此时已经存在的数据库。

```
hive> SHOW DATABASES;
OK
default
financials
human_resources
```

查询以字母 h 开头、以其他字符结尾(即 .* 部分含义)的数据库。

```
hive> SHOW DATABASES LIKE 'h.*';
OK
human_resources
```

2.2 数据库的修改与删除

1. 修改数据库

用户可以使用 ALTER DATABASE 命令为某个数据库的 DBPROPERTIES 设置键-值对属性值,用来描述这个数据库的属性信息。数据库的其他元数据信息都是不可以更改的,包括数据库名和数据库所在的目录位置。

```
hive>ALTER DATABASE financials SET DBPROPERTIES('edited-by'='Joe Dba');
```

数据库属性是无法删除或者"重置"的。

2. 删除数据库

以删除 financial 数据库为例学习如何删除数据库。

```
hive>DROP DATABASE IF EXISTS financials;
```

IF EXISTS 子句是可选的,加了这个子句,可以避免因数据库 financials 不存在而抛出警告信息。

默认情况下,Hive 不允许用户删除一个包含有表的数据库。如果用户想要删除一个有一张或多张表的数据库,那么可以先删除数据库中的表,然后再删除数据库,也可以在删除命令的最后面加上关键字 CASCADE,这样可以让 Hive 自行先删除数据库中的表。

```
hive>DROP DATABASE IF EXISTS financials CASCADE;
```

如果使用的是 RESTRICT 这个关键字,而不是 CASCADE 这个关键字,那么就和默认情况一样。也就是说，如果想删除数据库，就必须先要删除该数据库中的所有表。

如果某个数据库被删除了，那么其对应的目录也同时会被删除。

2.3 创 建 表

CREATE TABLE 语句遵从 SQL 语法惯例，但是该语句在 Hive 中具有显著的功能扩展，使其可以具有更好的灵活性。例如，可以定义表的数据文件存储在什么位置、使用什么样的存储格式等。

下面创建的这张表结构适用于 1.4.2 小节学习的"集合数据类型"中所声明的 employees 表。创建这张表有两种方法。

第一种方法：

```
hive>use mydb;
hive>CREATE TABLE IF NOT EXISTS employees(
    >name STRING COMMENT 'Employee name',
    >salary FLOAT COMMENT 'Employee salary',
    >subordinates ARRAY<STRING>COMMENT 'Names of subordinates',
    >deductions MAP<STRING, FLOAT>COMMENT 'Keys are deductions name, values are
      percentages',
    >address STRUCT< street: STRING, city: STRING, state: STRING, zip: STRING >
      COMMENT 'Home address')
    >COMMENT 'Description of the table'
    >ROW FORMAT DELIMITED
    >FIELDS TERMINATED BY '\t'
    >LOCATION '/user/hive/warehouse/mydb.db/employees';
```

第二种方法：

```
hive>CREATE TABLE IF NOT EXISTS mydb.employees(
    >name STRING COMMENT 'Employee name',
    >salary FLOAT COMMENT 'Employee salary',
    >subordinates ARRAY<STRING>COMMENT 'Names of subordinates',
    >deductions MAP<STRING, FLOAT>COMMENT 'Keys are deductions name, values are
      percentages',
    >address STRUCT<street: STRING, city: STRING, state: STRING, zip: STRING>
      COMMENT 'Home address')
    >COMMENT 'Description of the table'
    >ROW FORMAT DELIMITED
    >FIELDS TERMINATED BY '\t'
    >LOCATION '/user/hive/warehouse/mydb.db/employees';
```

选项 IF NOT EXISTS 的作用已经在 2.1 节介绍过了。对于已经存在的表，Hive 会忽略掉后面的执行语句，而且不会有任何提示。在那些第一次执行时需要创建表的脚本中，这个子句是非常有用的。

用户可以通过关键字 COMMENT 在字段类型后为每个字段增加一个注释。和数据库一样，用户也可以根据需求对表添加一个注释。

ROW FORMAT DELIMITED 这一行在 Hive 语法中表示的是行的格式。FIELDS

TERMINATED BY '\t'这一行定义了数据以'\t'分隔符来分隔,当然也可以定义以其他的分隔符来分隔数据。

从创建表语句的最后一行可以看出,可以根据情况为表中的数据指定一个存储路径(和元数据截然不同的是,元数据总是会保存这个路径)。在上面这个例子中,Hive 将会使用默认的路径/user/hive/warehouse/mydb.db/employees。其中,/user/hive/warehouse 是默认的"数据仓库"路径地址,mydb.db 是数据库目录,employees 是表目录。

默认情况下,Hive 总是将创建的表的目录放置在这个表所属的数据库目录之后。但是,default 数据库例外,其在/user/hive/warehouse 下并没有对应的数据库目录,default 数据库中的表目录会直接位于/user/hive/warehouse 目录之后(除非用户明确指定为其他路径)。

提示:在所有的数据库相关的命令中,都可以使用 SCHEMA 关键字来替代关键字 TABLE。

2.3.1 管理表

通常所创建的表都是管理表(也称为内部表),Hive 会或多或少地控制着管理表中数据的生命周期。每个管理表在 Hive 中都有一个对应 HDFS 上的存储目录,默认情况下会将这些表的数据存储在由配置项 hive.metastore.warehouse.dir(例如,/user/hive/warehouse)所定义目录的子目录下。

创建管理表的操作包含两个步骤:表创建和数据加载(这两个过程可以在同一条语句中完成)。在数据加载过程中,实际数据会移动到数据仓库目录,之后的数据访问过程将会直接在数据仓库目录中完成。当删除一个管理表时,Hive 也会删除这个表中的数据。

2.3.2 外部表

管理表不方便和其他工作共享数据。例如,假设有一份由 Pig 或者其他工具创建并且主要由这一工具使用的数据,还想使用 Hive 在这份数据上执行一些查询操作,可是并没有给予 Hive 对数据的所有权,可以创建一个外部表指向这份数据(并不需要对其具有所有权)。从内部表、外部表的特点来看,在删除内部表时,Hive 将会把属于表的元数据和数据全部删掉;而删除外部表时,Hive 仅仅删除外部表的元数据,数据是不会删除的。

创建外部表是通过指定 EXTERNAL 关键字来实现的。外部表对应的文件存储在 LOCATION 指定的 HDFS 路径下,并不会移动到数据仓库目录,在向该目录添加新文件的同时,该表也会读取到该文件(文件格式必须跟表定义的一致),删除外部表时并不会删除 LOCATION 指定目录下的文件。

下面创建一个外部表,其可以读取所有位于/data/stocks 目录下的以逗号分隔的数据。

```
hive>CREATE EXTERNAL TABLE IF NOT EXISTS stocks(
    > exchanges STRING,
    > symbol STRING,
    > ymd STRING,
    > price_open FLOAT,
    > price_high FLOAT,
```

```
          > price_low FLOAT,
          > price_close FLOAT,
          > volume INT,
          > price_adj_close FLOAT)
          > ROW FORMAT DELIMITED
          > FIELDS TERMINATED BY ','
          > LOCATION '/data/stocks';
```

管理表和外部表之间的差异并不大。即使对于管理表,用户也是可以知道数据是位于哪个路径下的,因此用户也可以使用其他工具(如 Hadoop 的 dfs 命令等)来修改、删除管理表所在的路径目录下的数据。从严格意义上来说,Hive 管理着这些目录和文件,但是并不具有对它们的完全控制权限。当然,管理表和外部表还有一些小小的区别,有些 HiveQL 语法结构不适用于外部表。

2.3.3 查看表结构

在 Hive 中,用 show tables 来查询有哪些表,用 describe tablename 来查看表结构,还可以用 show create table tablename 来查看建表源代码及其表结构。

用 show tables 来查询有哪些表。

```
hive> show databases;
OK
default
human_resources
mydb
Time taken: 0.091 seconds, Fetched: 3 row(s)
hive> use mydb;
OK
Time taken: 0.048 seconds
hive> show tables;
OK
employees
stocks
```

用 describe tablename 来查看表结构。

```
hive> describe employees;
OK
name                    string                  Employee name
salary                  float                   Employee salary
subordinates            array<string>           Names of subordinates
deductions              map<string,float>       Keys are deductions name,values a
re percentages
address                 struct<street:string,city:string,state:string,zip:string>
Home address
Time taken: 0.169 seconds, Fetched: 5 row(s)
```

用 show create table tablename 来查看建表源代码及其表结构。

```
hive> show create table employees;
OK
CREATE TABLE 'employees'(
  'name' string COMMENT 'Employee name',
  'salary' float COMMENT 'Employee salary',
  'subordinates' array<string> COMMENT 'Names of subordinates',
  'deductions' map<string,float> COMMENT 'Keys are deductions name,values are percentages',
  'address' struct<street:string,city:string,state:string,zip:string> COMMENT 'Home address')
COMMENT 'Description of the table'
ROW FORMAT DELIMITED
```

```
    FIELDS TERMINATED BY '\t'
STORED AS INPUTFORMAT
    'org.apache.hadoop.mapred.TextInputFormat'
OUTPUTFORMAT
    'org.apache.hadoop.hive.ql.io.HiveIgnoreKeyTextOutputFormat'
LOCATION
    'hdfs://master:9000/user/hive/warehouse/mydb.db/employees'
TBLPROPERTIES (
    'transient_lastDdlTime'='1458629157')
Time taken: 0.645 seconds, Fetched: 17 row(s)
```

2.4 修 改 表

大多数的表属性可以通过 ALTER TABLE 语句进行修改,该操作会修改元数据,但是不会修改数据本身。下面以 employees 表为例来学习对表的一些具体的修改操作。

1. 表的重命名

使用以下语句可以将表 employees 重命名为 emp。

```
hive>ALTER TABLE employees RENAME TO emp;
```

2. 修改列信息

用户可以对某个字段进行重命名,修改其类型和注释。例如,用以下语句将表 employees 的列 name 重命名为 ename,把注释改为 Employees name,然后将 salary 的数据类型 FLOAT 改为 DOUBLE。

```
hive>ALTER TABLE employees CHANGE COLUMN name ename STRING
    >COMMENT 'Employees name';
hive>ALTER TABLE employees CHANGE COLUMN salary salary DOUBLE;
```

这个命令只会修改元数据信息。其中,关键字 COLUMN 和 COMMENT 子句是可选的。

查看修改过后的列信息。

```
hive> describe employees;
OK
ename                   string                  Employees name
salary                  double                  Employee salary
```

如果需要把修改的字段移到某个字段的后面,可以使用 AFTER other_column 子句;如果想要移到第一行,可以使用 FIRST 代替 AFTER other_column 子句。如果用户移动的是字段,那么数据应与新的模式匹配,或者通过其他某些方法修改数据使其能够和模式匹配。建议不要改变字段的位置。

3. 增加列

在表 employees 中增加一列列名为 dept、类型为 STRING、列注释为 Department name 的新列。

```
hive>ALTER TABLE employees ADD COLUMNS(
    >dept STRING COMMENT 'Department name');
```

其中,COMMENT 子句是可选的。

查看新增加的列。

```
hive> describe employees;
OK
ename                   string                  Employees name
salary                  double                  Employee salary
subordinates            array<string>           Names of subordinates
deductions              map<string,float>       Keys are deductions name,values
are percentages
address                 struct<street:string,city:string,state:string,zip:string
>            Home address
dept                    string                  Department name
```

如果新增的字段中有某个或多个字段位置是错误的，那么需要使用 ALTER COLUMN tablename CHANGE COLUMN 语句，逐一将字段调整到正确的位置。

4．删除或替换列

在表 employees 中，删除列并重新指定了新的字段 name。

```
hive>ALTER TABLE employees REPLACE COLUMNS(
    >name STRING COMMENT 'Employee name');
```

这个语句删除了所有的列，并且从之前的表定义的模式中只保留了 name。因为是 ALTER 语句，所以只有表的元数据信息改变了。REPLACE 语句只能用于使用了如下两种内置 SerDe 模块的表：DynamicSerDe 或 MetadataTypedColumnsetSerDe。

5．修改表的属性

（1）修改表的属性

用户可以增加附加的表属性或者修改已经存在的属性，但是不能删除属性。修改属性的方法如下：

```
hive>ALTER TABLE employees SET TBLPROPERTIES(
    >'notes'='The column is always NULL');
```

（2）修改表的存储属性

将一张表的存储格式修改为 SEQUENCEFILE。

```
hive>ALTER TABLE employees
    >SET FILEFORMAT SEQUENCEFILE;
```

修改表 stocks 的存储属性。

```
hive>ALTER TABLE stocks
    >CLUSTERED BY(exchanges, symbol)
    >SORTED BY(symbol)
    >INTO 48 BUCKETS;
```

CLUSTERED BY 子句和 INTO…BUCKETS 子句是必需的，SORTED BY 子句是可选的。

2.5 删 除 表

删除表是一个比较简单的操作，因为 Hive 支持和 SQL 中 DROP TABLE 命令类似的操作。例如，删除 employees1。

```
hive>DROP TABLE employees1;
```

对于管理表(内部表),表的元数据信息和表内的数据都会被删除;对于外部表,表的元数据信息会被删除,但是表中的数据不会被删除。

提示:如果用户开启了 Hadoop 回收站功能(此功能默认情况下是关闭的),那么数据将会被转移到用户在分布式文件系统中的用户根目录下的 .Trash 目录下,也就是 HDFS 中的 /user/$USER/.Trash 目录。如果想开启这个功能,只需要将配置属性 fs.trash.interval 的值设置为一个合理的正整数。这个值是"回收站检查点"间的时间间隔,单位是分钟。如果设置的值是 1440,那么就表示 24 小时。不过并不能保证所有的分布式系统以及所有版本都支持这个功能。如果用户不小心删除了一张存储着重要数据的管理表,那么可以先重建表,然后重建所需要的分区,再从 .Trash 目录中将误删的文件移动到正确的文件目录下(使用文件系统命令)来重新存储数据。

2.6 分区表

Hive Select 查询一般会扫描整个表,会消耗很多时间做没必要的工作,有时只需要扫描表中关心的一部分数据,因此建表时引入了分区表(partition)的概念。分区表具有重要的性能优势,可以有多种形式。通常使用分区来水平分散压力,将数据从物理上转移到距使用最频繁的用户更近的地方,实现优化的目的;另外,分区表还可以将数据以一种符合逻辑的方式进行组织,如分层存储。

在 Hive 中创建分区表没有复杂的分区类型(范围分区、列表分区、hash 分区、混合分区等)。分区列也不是表中的一个实际的字段,而是一个或者多个伪列,意思是说在表的数据文件中实际上并不保存分区列的信息与数据。

创建 employees 分区表。

```
hive>CREATE TABLE employees(
    >name STRING,
    >salary FLOAT,
    >subordinates ARRAY<STRING>,
    >deductions MAP<STRING, FLOAT>,
    >address STRUCT<street:STRING, city:STRING, state:STRING, zip:INT>)
    >PARTITIONED BY(country STRING, state STRING)
    >ROW FORMAT DELIMITED
    >FIELDS TERMINATED BY '\t'
    >LOCATION '/user/hive/warehouse/mydb.db/employees';
```

注意:语句中使用 PARTITIONED 关键字,以 country 和 state 两个字段作为分区列。

通常情况下需要先创建好分区,然后才能使用该分区。例如:

```
hive>ALTER TABLE employees ADD PARTITION(country='China',state='Beijing');
```

这样就创建好了一个分区,这时会看到 Hive 在 HDFS 存储中创建了一个相应的文件夹。

```
[zkpk@master ~]$ hadoop fs -ls /user/hive/warehouse/mydb.db/employees/country=Ch
ina
SLF4J: Class path contains multiple SLF4J bindings.
SLF4J: Found binding in [jar:file:/home/zkpk/hadoop-2.5.2/share/hadoop/common/li
b/slf4j-log4j12-1.7.5.jar!/org/slf4j/impl/StaticLoggerBinder.class]
SLF4J: Found binding in [jar:file:/home/zkpk/hbase-1.1.2/lib/slf4j-log4j12-1.7.5
.jar!/org/slf4j/impl/StaticLoggerBinder.class]
SLF4J: See http://www.slf4j.org/codes.html#multiple_bindings for an explanation.
SLF4J: Actual binding is of type [org.slf4j.impl.Log4jLoggerFactory]
16/03/23 01:25:27 WARN util.NativeCodeLoader: Unable to load native-hadoop libra
ry for your platform... using builtin-java classes where applicable
Found 1 items
drwxr-xr-x   - zkpk supergroup          0 2016-03-23 01:21 /user/hive/warehouse/
mydb.db/employees/country=China/state=Beijing
```

每一个分区都会有一个独立的文件夹，存放的是该分区所有的数据文件。其中，country 是主层次，state 是副层次。所有 country 不同的分区都会在/user/hive/warehouse/mydb.db/employees 下，如果 country='China'，则 state 不同的分区都会在/user/hive/warehouse/mydb.db/employees/country= China 下。

```
[zkpk@master mapreduce]$ hadoop fs -ls /user/hive/warehouse/mydb.db/employees
SLF4J: Class path contains multiple SLF4J bindings.
SLF4J: Found binding in [jar:file:/home/zkpk/hadoop-2.5.2/share/hadoop/common/lib
/slf4j-log4j12-1.7.5.jar!/org/slf4j/impl/StaticLoggerBinder.class]
SLF4J: Found binding in [jar:file:/home/zkpk/hbase-1.1.2/lib/slf4j-log4j12-1.7.5.
jar!/org/slf4j/impl/StaticLoggerBinder.class]
SLF4J: See http://www.slf4j.org/codes.html#multiple_bindings for an explanation.
SLF4J: Actual binding is of type [org.slf4j.impl.Log4jLoggerFactory]
16/03/23 01:33:38 WARN util.NativeCodeLoader: Unable to load native-hadoop librar
y for your platform... using builtin-java classes where applicable
Found 2 items
drwxr-xr-x   - zkpk supergroup          0 2016-03-23 01:21 /user/hive/warehouse/m
ydb.db/employees/country=China
drwxr-xr-x   - zkpk supergroup          0 2016-03-23 01:33 /user/hive/warehouse/m
ydb.db/employees/country=US
```

如果当前表中存在很多的分区，而用户只想查看是否存在某个特定分区键的分区，用户还可以在下面这个命令上增加一个或者多个特定分区字段值的 PARTITION 子句，进行过滤查询。

```
hive> SHOW PARTITIONS employees PARTITION(country='China');
OK
country=China/state=Beijing
Time taken: 0.362 seconds, Fetched: 1 row(s)

hive> SHOW PARTITIONS employees PARTITION(country='US',state='Il');
OK
country=US/state=Il
Time taken: 0.315 seconds, Fetched: 1 row(s)
```

注意：因为分区列的值要转化为文件夹的存储路径，所以如果分区列的值中包含特殊值，如'%'、':'、'/'、'#'等，它将会被使用％加上 2 字节的 ASCII 码进行转义。

2.6.1 外部分区表

跟内部表分区一样，外部表也可以使用分区。实际上，这是管理大型生产数据集最常见的情况。外部表分区给用户提供了一个可以和其他工具共享数据的方式，同时也可以优化查询性能。举一个非常具有说服力的例子——日志文件分析：大多数系统会使用一个标准的日志信息格式，而时间是每条日志的基本信息，用时间来进行分区对表的存储非常有效。按照设定将日志数据按照天进行划分的数据尺寸合适，而且按天这个粒度进行查询速度足够快。按日期创建外部分区表。

```
hive>CREATE EXTERNAL TABLE IF NOT EXISTS log_messages(
    >hms INT,
    >severity STRING,
    >server STRING,
    >process_id INT,
    >message STRING)
    >PARTITIONED BY(year INT, month INT, day INT)
    >ROW FORMAT DELIMITED
    >FIELDS TERMINATED BY '\t';
```

在创建某些表时,有些要求写 LOCATION 语句,对于外部分区表则没有这样的要求。

ALTER TABLE 语句可以单独增加分区。语句中需要为每一个分区键指定一个值,如上面创建的例子中需要为 year、month 和 day 这 3 个分区键都指定值。例如,增加一个 2016 年 3 月 23 日的分区。

```
hive>ALTER TABLE log_messages ADD PARTITION(year=2016,month=3,day=23)
    >LOCATION '/user/hive/warehouse/mydb.db/log_messages/2016/03/23';
```

和分区管理表一样,通过 SHOW PARTITIONS 命令可以查看一个外部表的分区。

```
hive> SHOW PARTITIONS log_messages;
OK
year=2016/month=3/day=23
Time taken: 0.292 seconds, Fetched: 1 row(s)
```

还可以用 DESCRIBE tablename 和 DESCRIBE EXTENDED tablename 这两种方式来查看创建的分区。

```
hive> describe  log_messages;
OK
hms                     int
severity                string
server                  string
process_id              int
message                 string
year                    int
month                   int
day                     int

# Partition Information
# col_name              data_type               comment

year                    int
month                   int
day                     int
Time taken: 0.123 seconds, Fetched: 15 row(s)
```

执行 DESCRIBE EXTENDED tablename 命令,会展示更加详细的表结构,但是它还缺少一个非常重要的信息,那就是分区数据实际存在的路径,可以通过如下方式查看。

```
hive>DESCRIBE EXTENDED log_messages PARTITION(year=2016,month=3,day=23);
```

分区外部表具有非常多的优点,如方便逻辑数据管理、实现高性能的查询等。

另外,ALTER TABLE tablename ADD PARTITION 语句并非只能对外部表使用。对于内部表,当有分区数据不是由 LOAD 和 INSERT 语句产生时,用户同样可以使用这个命令指定分区路径。

注意：并非所有的表数据都是放在默认的 Hive"warehouse" 目录下；同时，当删除内部表时，这些数据不会被删除掉。

2.6.2 自定义表的存储格式

Hive 的默认存储格式是文本文件格式。存储格式可以通过可选的子句 STORED AS TEXTFILE 显示指定，同时用户还可以在创建表时指定各种各样的分隔符。重新展示例 1-4 创建过的 employees 表。

```
hive>CREATE TABLE employee(
    >name STRING,
    >salary FLOAT,
    >subordinates ARRAY<STRING>,
    >deductions MAP<STRING, FLOAT>,
    >address STRUCT<street:STRING,city:STRING,state:STRING,zip:INT>)
    >ROW FORMAT DELIMITED
    >FIELDS TERMINATED BY '\001'
    >COLLECTION ITEMS TERMINATED BY '\002'
    >MAY KEYS TERMINATED BY '\003'
    >LINES TERMINATED BY '\n'
    >STORED AS TEXTFILE;
```

TEXTFILE 意味着所有字段都使用字母、数字、字符编码，包括国际字符集，尽管 Hive 默认是使用不可见字符来作为分隔符的。使用 TEXTFILE 就意味着每一行被认为是一个单独的记录。

用户也可以将 TEXTFILE 替换为其他 Hive 所支持的内置文件格式，包括 SEQUENCEFILE 和 RCFILE，这两种内置文件格式都是使用二进制编码和压缩（可选的）来优化磁盘空间使用和 I/O 带宽性能的。

对于记录是如何被编码成文件的，以及列是如何被编码为记录的，Hive 指出了它们之间的不同。用户可以分别自定义这些行为。记录编码是通过 inputformat 对象来控制的（如 TEXTFILE 后面的 Java 代码的实现）。Hive 使用了一个名为 org.apache.hadoop.mapred.TextInputFormat 的 Java 类（编译后的模块）。如果用户不熟悉 Java，这种点分隔的命名语法表明了包的一个分层的树形命名空间，这个结构和 Java 代码的目录结构是对应的。TextInputFormat 是位于最顶层包 mapred 下的一个类。

记录的解析是由序列化器/反序列化器（或者编写 SerDe）来控制的。Hive 所使用的 SerDe 是另外一个 Java 类 org.apache.hadoop.hive.serde2.lazy.LazySimpleSerDe。为了保持完整性，Hive 还使用一个叫做 outputformat 的对象来将查询的结果写入文件或者输出到控制台。对于 TEXTFILE，用于输出的 Java 类名是 org.apache.hadoop.hive.ql.io.HiveIgnoreKeyTextOutputFormat。

当然，用户还可以指定第三方的输入、输出格式和 SerDe，这个功能允许用户自定义 Hive 本身不支持的其他广泛的文件格式。

注意：Hive 使用 inputformat 对象将输入流分隔成记录，然后使用 outputformat 对象将记录格式化为输出流，再使用 SerDe 在读数据时将记录解析为列，在写数据时将列编码成记录。

2.6.3 增加、修改和删除分区表

正如在 2.6.1 小节所见到的,ALTER TABLE tablename ADD PARTITION...语句用于为外部表增加一个分区。下面,增加可提供的可选项,然后多次重复前面的分区路径。

```
hive> ALTER TABLE log_messages ADD IF NOT EXISTS
    > PARTITION(year=2016,month=3,day=24)LOCATION '/user/hive/warehouse/mydb.db/
log_messages/2016/03/24'
    > PARTITION(year=2016,month=3,day=25)LOCATION '/user/hive/warehouse/mydb.db/
log_messages/2016/03/25';
```

当用户使用 Hive v0.8.0 及其之后的版本时,在同一个查询中可以同时增加多个分区。IF NOT EXISTS 子句和之前的一样,是个可选子句,且含义不变。至于如何查看新添加的表分区,在 2.6.1 小节也学习过了,这里就不多做介绍了。

另外,还可以通过高效地移动位置来修改某个分区的路径。

```
ALTER TABLE log_messages PARTITION(year=2016,month=3,day=25)SET LOCATION '新的
路径';
```

删除表 log_messages 中的分区(year=2016,month=3,day=25):

```
hive>ALTER TABLE log_messages DROP IF EXISTS
    PARTITION(year=2016,month=3,day=25);
```

2.7 桶 表

Hive 引入的 partition 和 bucket 概念,中文翻译分别为分区和桶。这两个概念都是把数据划分成块。对于每一张表或者分区,Hive 可以进一步组织成桶,也就是说桶是更为细粒度的数据范围划分,如图 2-1 所示。分区是粗粒度的划分,桶是细粒度的划分。

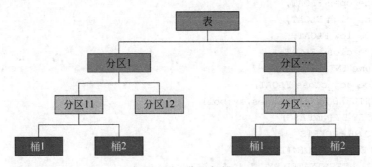

图 2-1 分区与桶

Hive 可以针对某一列进行桶的组织。Hive 采用对列哈希(Hash)值,然后除以桶的个数求余的方式决定该条记录存放在哪个桶中。

桶表的优点:

(1) 获得更高的查询处理效率。桶为表加上了额外的结构,Hive 在处理某些查询时能利用这个结构。具体而言,两个在相同列上划分了桶的表可以使用 Map 端实现高效的连

接，如 JOIN 操作。JOIN 操作有一个相同列的两个表，如果对这两个表都进行了桶操作，那么将保存相同列值的桶进行 JOIN 操作，就可以大大地减少 JOIN 的数据量。

（2）取样更高效。在处理大规模数据集时，在开发和修改查询的阶段，如果能在数据集的一小部分数据上运行查询，会带来很多方便。使用桶表可以让查询发生在小范围的数据中以提高效率。

下面通过一个例子介绍在 Hive 中如何创建桶表。

```
hive>set hive.enforce.bucketing=true;
hive>CREATE TABLE bucketed_users(
    >id INT,
    >name STRING)
    >CLUSTERED BY(id)
    >SORTED BY(name)
    >INTO 4 BUCKETS
    >ROW FORMAT DELIMITED
    >FIELDS TERMINATED BY '\t'
    >STORED AS TEXTFILE
    >LOCATION '/user/hive/warehouse/mydb.db/bucketed_users';
```

其中，CLUSTERED BY 子句用于指定划分桶所用的列和要划分的桶的个数，其他的与创建表的思想是一样的，之后的查询操作和分区表的一样。CLUSTERED BY … INTO … BUCKETS 子句后面还可以接一个可选的 SORTED BY …子句，用于优化某些特定类型的查询。

下面再给出一个例子，以加深大家对它的印象。

```
hive>CREATE EXTERNAL TABLE IF NOT EXISTS stock(
    >exchanges STRING,
    >symbol STRING,
    >ymd STRING,
    >price_open FLOAT,
    >price_high FLOAT,
    >price_low FLOAT,
    >price_close FLOAT,
    >volume INT,
    >price_adj_close FLOAT)
    >CLUSTERED BY(exchanges,symbol)
    >SORTED BY(ymd ASC)
    >INTO 96 BUCKETS
    >ROW FORMAT DELIMITED
    >FIELDS TERMINATED BY  ','
    >LOCATION  '/data/stock';
```

本章小结

本章的学习主要是以实例展开的。

（1）学习了创建、查看、删除和修改数据库。需要注意的是，在修改数据库信息时，只能修改数据库的属性信息,不能修改数据库的元数据信息,数据库名和数据库所在的目录位置也不能修改。

(2) 学习了什么是管理表、外部表,以及它们的创建和查看表结构。学习了修改表的一些操作,以及如何删除表。

(3) 学习了分区表、桶表,区别分区表和桶表。

习 题

1. 选择题

(1) 如果用户没有显示地指定数据库,那么将会使用默认的数据库(　　)。

  A. public   B. default   C. private   D. protect

(2) 当删除一个(　　)时,Hive 也会删除这个表中的数据。

  A. 内部表   B. 外部表   C. 分区表   D. 桶表

(3) (　　)是在创建外部表时用到的关键字。

  A. SHOW   B. DESCRIBE   C. EXTERNAL   D. DROP

(4) 以下选项中,不属于分区表的分区类型的是(　　)。

  A. 范围分区   B. Hash 分区   C. 时间分区   D. 列表分区

(5) 下列不属于文件存储格式的是(　　)。

  A. SEQUENCEFILE   B. TEXTFILE

  C. RCFILE   D. FILE

2. 问答题

(1) 数据库默认的路径是什么?表的默认路径又是什么?

(2) 管理表和外部表的区别有哪些?

(3) 在对表进行修改时,移动表中的字段位置时,表中的数据会有哪些变化?遇到这种问题时,该怎样解决?

(4) 思考分区表是不是分区越多越好。

(5) 用自己的话描述图 2-1,并且指出桶表的优点。

第 3 章

HiveQL 数据操作

本章摘要

在第 2 章中学习了管理表、外部表、分区表和桶表的概念,并且学习了创建、修改和删除表。在本章中将学习如何对 Hive 表中的数据进行操作。

首先,将学习如何把数据加载到表中,以及如何把表中的数据导出到指定的位置;其次,作为本章最重要的部分,将要学习一些数据查询语句,包括简单的查询语句和复杂的查询语句,因为好多用户需求需要通过查询语句得出结果;最后,简单介绍抽样查询。

3.1 数据加载与导出

本节主要学习如何往表中加载数据,以及怎么导出数据。数据加载的几种主要方式:从本地系统中导入数据到 Hive 表中、从 HDFS 上导入数据到 Hive 表中、通过查询语句向 Hive 表中导入数据,以及动态分区插入数据;数据导出包括如何导出数据和导到何处。

3.1.1 数据加载

1. 从本地系统中导入数据到 Hive 表中

实例:

hive>LOAD DATA LOCAL INPATH '/home/zkpk/empmessages' INTO TABLE emp_messages;

本例是把本地文件 empmessages 中的数据导入创建的 Hive 表 emp_messages 中。通过这个实例,整理出通用的语句格式是:

LOAD DATA LOCAL INPATH '/本地文件路径' INTO TABLE tablename;

注意:关键字 LOCAL 不能漏掉。如果忘记此关键字,则默认从 HDFS 上去寻找所需要的数据文件路径。

2. 从 HDFS 上导入数据到 Hive 表中

实例:

[zkpk@master ~]$hadoop fs -put /home/zkpk/empmessages /data/emp_messages;

由于已创建的管理表就在 HDFS 上,所以只需要将文件 put 到管理表所在的目录就可以完成数据导入。通过这个实例,整理出来通用的语句格式是:

hadoop fs -put '/数据所存放的本地路径' '/创建表时存放表的路径'

3. 通过查询语句向 Hive 表中导入数据

通过查询语句向一张 Hive 空表中导入数据的实例。

```
hive> INSERT INTO TABLE emp_messages SELECT * FROM old_emp_messages;
```

如该例所示，首次向表中插入数据时，语句为 INSERT INTO TABLE tablename SELECT…INTO…。

查看此时表中的数据，验证执行结果。

```
hive> select * from emp_messages;
OK
1       zhangsan    23      man
2       lisi        21      woman
3       wangbin     24      man
4       liji        20      woman
```

使用 OVERWRITE 关键字时，可以覆盖目标表中原来相同 partition 中的所有数据，如果目标表中没有 partition，则覆盖整个表。

```
hive> INSERT OVERWRITE TABLE emp_messages SELECT * FROM old_emp_messages;
```

查看此时表中的数据，验证执行结果。

```
hive> select * from emp_messages;
OK
1       zhouxiao    20      woman
2       zhaoliun    26      man
3       zhangjiu    26      man
```

4. 动态分区插入数据

所谓动态分区，也称作动态分区插入，指的是插入目标表时仅指定分区字段，不指定分区值，分区值是从原始表中取得的。静态分区和动态分区的区别在于导入数据时，是手动输入分区名称，还是通过数据来判断数据分区。如果一次插入上百上千个分区中，只写插入的代码就很多，这种场景就适合使用动态分区插入功能。

默认情况下，Hive 是支持动态分区插入的，但是并没有开启。开启后，默认是以"严格"模式执行：要求至少有一列分区字段是静态的。这有助于阻止因设计错误导致查询产生大量的分区。表 3-1 描述了动态分区相关的属性设置。

表 3-1 动态分区属性

属 性 名 称	默认值	描　　述
hive.exec.dynamic.partition	false	设置成 true，表示开启动态分区功能
hive.exec.dynamic.partition.mode	strict	设置成 nonstrict，表示允许所有分区都是动态的
hive.exec.max.dynamic.partitions.pernode	100	每个 Mapper 或 Reducer 可以创建的最大动态分区个数。如果某个 Mapper 或 Reducer 尝试创建大于这个值的分区，则会抛出一个致命的错误信息
hive.exec.max.dynamic.partitions	1000	一个动态分区创建语句可以创建的最大动态分区个数。如果超过这个值，则会抛出一个致命错误信息
hive.exec.max.created.files	100000	全局可以创建的最大文件个数。有一个 Hadoop 计数器会跟踪记录创建了多少个文件，如果超过这个值，则会抛出一个致命错误信息

在学习动态分区插入数据前,先学习静态分区插入数据。所谓的静态分区插入数据就是在写插入语句时,分区的值为一个确定的值,通过如下的例子可以加深认识。

```
hive>INSERT OVERWRITE TABLE testpar PARTITION(days='0328')
    >SELECT * FROM test t WHERE t.day='0328';
```

接下来,查看一下插入的结果。

```
hive> select * from testpar;
OK
1       john    24      0328    0328
2       joe     25      0328    0328
3       lis     28      0328    0328
```

动态分区,顾名思义,就是在分区的字段值不确定的情况下进行数据插入操作。

```
hive>INSERT OVERWRITE TABLE testpar PARTITION(days)
    >SELECT * FROM test;
```

注意:如果分区是可以确定的,建议用静态分区的方式。不要用动态分区,因为动态分区的值是在 reduce 运行阶段确定的,也就是会把所有的记录 distribute by。可想而知,表记录非常大的情况下,只有一个 reduce 处理,后果是不可想象的。然而,静态分区在编译阶段已经确定,不需要 reduce 处理。

3.1.2 数据导出

数据导出可以分为:导出到本地文件系统中;导出到 HDFS 文件中;从一张表导出到另一张表中。

1. 把数据导出到本地文件系统中

```
hive>INSERT OVERWRITE LOCAL DIRECTORY '/home/zkpk/test'
    >SELECT * FROM test;
```

执行完上面的语句,在本地文件系统中查看有没有导出数据的文件。

```
[zkpk@master ~]$ cd /home/zkpk/test
[zkpk@master test]$ ll
total 4
-rw-r--r--. 1 zkpk zkpk 75 Mar 27 20:42 000000_0
```

2. 把数据导出到 HDFS 文件中

```
hive>INSERT OVERWRITE DIRECTORY '/data/test_1'
    >SELECT * FROM test;
```

查看结果如下:

```
[zkpk@master test]$ hadoop fs -ls /data/test_1
SLF4J: Class path contains multiple SLF4J bindings.
SLF4J: Found binding in [jar:file:/home/zkpk/hadoop-2.5.2/share/hadoop/common/lib/slf4j-log4j12-1.7.5.jar!/org/slf4j/impl/StaticLoggerBinder.class]
SLF4J: Found binding in [jar:file:/home/zkpk/hbase-1.1.2/lib/slf4j-log4j12-1.7.5.jar!/org/slf4j/impl/StaticLoggerBinder.class]
SLF4J: See http://www.slf4j.org/codes.html#multiple_bindings for an explanation.
SLF4J: Actual binding is of type [org.slf4j.impl.Log4jLoggerFactory]
16/03/27 20:51:38 WARN util.NativeCodeLoader: Unable to load native-hadoop library for your platform... using builtin-java classes where applicable
Found 1 items
-rw-r--r--   1 zkpk supergroup         75 2016-03-27 20:50 /data/test_1/000000_0
```

3. 从一张表导出到另一张表中

```
hive> INSERT INTO TABLE test_1
    > SELECT * FROM test;
```

查看结果如下：

```
hive> select * from test_1;
OK
1    john     24    0328
2    joe      25    0328
3    lis      28    0328
4    zhali    25    0329
5    wangw    23    0329
```

如果数据文件恰好是用户所需要的格式，那么只需要把数据复制到目标路径下。

```
hadoop fs -cp /source_path /target_path
```

3.2 数据查询

在本节中，通过搜狗搜索日志分析系统案例来介绍一些常用的数据查询语句。
创建表：

```
hive> CREATE TABLE sougou_20111230(
    > logdate STRING,
    > uid STRING,
    > keyword STRING,
    > rank INT,
    > searchorder INT,
    > url STRING)
    > ROW FORMAT DELIMITED
    > FIELDS TERMINATED BY '\t'
    > LOCATION '/data/sogou_20111230';

hive> CREATE TABLE sougou_old_20111230(
    > logdate STRING,
    > uid STRING,
    > keyword STRING,
    > rank INT,
    > searchorder INT,
    > url STRING)
    > ROW FORMAT DELIMITED
    > FIELDS TERMINATED BY '\t'
    > LOCATION '/data/sogou_old_20111230';
```

3.2.1 SELECT … FROM 语句

SELECT…FROM 语句和 MySQL 中的语法是一致的，SELECT 是 SQL 中的投影算子，FROM 子句标识了从哪个表、视图或嵌套查询中选择记录。

查询表 sougou_20111230 中的所有字段信息。

```
hive> SELECT * FROM sougou_20111230;
OK
20111230000009  698956eb07815439fe5f46e9a4503997       youku         1    1    http://www.youku.com/
20111230000014  f31f594bd1f3147298bd952ba35de84d       12306.cn      1    1    http://www.1230
6.cn/
20111230000018  596444b8c02b7b30c11273d5bbb88741       pissing videos 1   1    http://lakery.c
om/8y18-girl-pissing-in-mouth?nearest
20111230000019  11e2e89dbf484ed187e73cbeaf1e0084       www.june9.info@16  1  1   http://
r.baidu.com/1QCa0
20111230000019  63fd6f826a5f83d795f08778468d0e14       yunvxinjin    4    1    http://www.zvod
.net/zvoddianying/14028.html
```

一张表中的字段有可能非常多,当用户所需要查询的字段只需要少数几个时,把需要的字段列举出来即可。

```
hive> SELECT date,uid,keyword FROM sougou_20111230;
OK
20111230000009  698956eb07815439fe5f46e9a4503997       youku
20111230000014  f31f594bd1f3147298bd952ba35de84d       12306.cn
20111230000018  596444b8c02b7b30c11273d5bbb88741       pissing videos
20111230000019  11e2e89dbf484ed187e73cbeaf1e0084       www.june9.info@16
20111230000019  63fd6f826a5f83d795f08778468d0e14       yunvxinjin
```

对于集合数据类型,引用集合数据类型中的元素有以下方式。

(1) 数组引用方式,其索引是从 0 开始的(和 Java 一样),语句如下:

SELECT 字段名[集合数据类型中的元素的位置] FROM tablename;

注意:引用一个不存在的元素将会返回 NULL。

(2) 为了引用一个 MAP 元素,用户还可以使用 ARRAY[…]语法,但是使用的键值是非整数索引。语句如下:

SELECT 字段名[集合数据类型中的元素内容] FROM tablename;

(3) 为了引用 STRUCT 中的一个元素,用户可以使用".",符号,类似于"表的别名.列名"。

SELECT 字段名.集合中的某个元素 FROM tablename;

FROM 子句在使用中还有以下用法和功能。

1. LIMIT 语句

往往典型的查询会返回多行数据,有时候不需要查询那么多行,这时候可以使用LIMIT 关键字来限制行数。例如,查询表 sougou_20111230 中 5 行数据。

```
hive>SELECT * FROM sougou_20111230 LIMIT 5;
```

结果如下:

```
OK
20111230000009  698956eb07815439fe5f46e9a4503997       youku     1   1   h
ttp://www.youku.com/
20111230000014  f31f594bd1f3147298bd952ba35de84d       12306.cn  1   1
http://www.12306.cn/
20111230000018  596444b8c02b7b30c11273d5bbb88741       pissing videos 1  1
http://lakery.com/8y18-girl-pissing-in-mouth?nearest
20111230000019  11e2e89dbf484ed187e73cbeaf1e0084       www.june9.info@16  1
1       http://r.baidu.com/1QCa0
20111230000019  63fd6f826a5f83d795f08778468d0e14       yunvxinjin  4  1
http://www.zvod.net/zvoddianying/14028.html
Time taken: 0.085 seconds, Fetched: 5 row(s)
```

查询表中 3 行数据。

```
hive> SELECT * FROM sougou_20111230 LIMIT 3;
```

结果如下：

```
OK
20111230000009  698956eb07815439fe5f46e9a4503997    youku   1       1       h
ttp://www.youku.com/
20111230000014  f31f594bd1f3147298bd952ba35de84d    12306.cn        1       1
http://www.12306.cn/
20111230000018  596444b8c02b7b30c11273d5bbb88741    pissing videos 1        1
http://lakery.com/8y18-girl-pissing-in-mouth?nearest
Time taken: 0.088 seconds, Fetched: 3 row(s)
```

2．别名

顾名思义，别名就是为了方便给表或者列另外起的名字。表的别名和列的别名可以分开用，也可以同时出现在一个查询语句里。

（1）分开使用的实例。

```
hive> SELECT date,uid,keyword FROM sougou_20111230 sougou;
OK
20111230000009  698956eb07815439fe5f46e9a4503997    youku
20111230000014  f31f594bd1f3147298bd952ba35de84d    12306.cn
20111230000018  596444b8c02b7b30c11273d5bbb88741    pissing videos
20111230000019  11e2e89dbf484ed187e73cbeaf1e0084    www.june9.info@16
20111230000019  63fd6f826a5f83d795f08778468d0e14    yunvxinjin
20111230000020  637b29b47fed3853e117aa7009a4b621    fdf
20111230000027  4a6f0d5cc0bcf16e32e74ae49663b60d    baidu
20111230000027  9c89762b968568aaa0bed63579088f8e    stocking videos
20111230000041  ec0363079f36254b12a5e30bdc070125    clearaudio basic
20111230000058  4c4c23ff94387248f4dc88166177058a    baidu
Time taken: 0.175 seconds, Fetched: 10 row(s)
```

（2）列别名使用的实例。

```
hive> SELECT distinct keyword AS key FROM sougou_20111230;
12306.cn
baidu
clearaudio basic
fdf
pissing videos
stocking videos
www.june9.info@16
youku
yunvxinjin
```

（3）同时使用的实例。

```
hive> SELECT sougou.logdate as ld,sougou.searchorder as so
    > FROM sougou_20111230 sougou limit 10;
hive> SELECT sougou.date,sougou.order FROM sougou_20111230 sougou;
OK
20111230000009  1
20111230000014  1
20111230000018  1
20111230000019  1
20111230000019  1
20111230000020  1
20111230000027  1
20111230000027  1
20111230000041  1
20111230000058  1
Time taken: 0.075 seconds, Fetched: 10 row(s)
```

3. 嵌套 SELECT 语句

别名嵌套在查询中非常实用，以搜狗日志数据为例。

```
hive>SELECT count(so.key)
    >FROM(
    >SELECT distinct keyword AS key
    >FROM sougou_20111230)so LIMIT 10;
```

4. CASE…WHEN…THEN 语句

CASE…WHEN…THEN 语句和 if 条件语句类似，用于处理单个列的查询。例如：

```
hive>SELECT keyword,rank,
    >CASE
    >WHEN rank<=2 THEN 'fast'
    >WHEN rank<=3 THEN 'middle'
    >WHEN rank<=4 THEN 'slow'
    >END AS bracket FROM sougou_20111230 LIMIT 10;
OK
youku               1       fast
12306.cn            1           fast
pissing videos 1            fast
www.june9.info@16           1       fast
yunvxinjin          4           slow
fdf                 1       fast
baidu               2       fast
stocking videos 3           middle
clearaudio basic            4       slow
baidu       1           fast
Time taken: 0.072 seconds, Fetched: 10 row(s)
```

3.2.2 WHERE 语句

SELECT 语句用于选取字段，WHERE 语句用于过滤条件，两者结合使用可以查找到符合过滤条件的记录。

WHERE 语句中使用谓词表达式表示条件，当谓词表达式计算结果为 true 时，相应的行将被保留并输出。有多个谓词表达式时，可以使用 AND 和 OR 相连接。

使用 AND 连接多个谓词的 WHERE 语句例子。

```
hive> SELECT * FROM sougou_20111230
    > WHERE rank=1 AND order=1;
OK
20111230000009  698956eb07815439fe5f46e9a4503997        youku       1       1       h
ttp://www.youku.com/
20111230000014  f31f594bd1f3147298bd952ba35de84d        12306.cn        1       1
http://www.12306.cn/
20111230000018  596444b8c02b7b30c11273d5bbb88741        pissing videos 1        1
http://lakery.com/8y18-girl-pissing-in-mouth?nearest
20111230000019  11e2e89dbf484ed187e73cbeaf1e0084        www.june9.info@16       1
1       http://r.baidu.com/1QCa0
20111230000020  637b29b47fed3853e117aa7009a4b621        fdf     1       1       h
ttp://www.163pan.com/files/70z00000j.html
20111230000058  4c4c23ff94387248f4dc88166177058a        baidu       1       1       h
ttp://www.baidu.com/
```

使用 OR 连接多个谓词的 WHERE 语句例子。

```
hive> SELECT * FROM sougou_20111230
    > WHERE rank=2 OR rank=3;
OK
20111230000027    4a6f0d5cc0bcf16e32e74ae49663b60d    baidu    2    1    h
ttp://site.baidu.com/
20111230000027    9c89762b968568aaa0bed63579088f8e    stocking videos 3    1
http://www.sexystockingtops.com/updates/update.html
```

1. 谓词操作符

谓词操作符如表 3-2 所描述，用于连接谓词表达式，用复杂的逻辑关系作为条件。这些操作符同样可以用于 JOIN...ON 和 HAVING 语句中。

表 3-2 谓词操作符

操 作 符	支持的数据类型	描 述
A=B	基本数据类型	如果 A 等于 B，则返回 TRUE，反之返回 FALSE
A<=>B	基本数据类型	如果 A 和 B 都为 NULL，则返回 TRUE，其他的和符号（=）操作符的结果一致，任一为 NULL，结果就为 NULL
A==B	没有	这个语法是不正确的。SQL 使用=
A<>B, A!=B	基本数据类型	如果 A 或者 B 为 NULL，则返回 NULL；如果 A 不等于 B，则返回 TRUE，否则返回 FALSE
A<B	基本数据类型	如果 A 或者 B 为 NULL，则返回 NULL；如果 A 小于 B，则返回 TRUE，否则返回 FALSE
A<=B	基本数据类型	如果 A 或者 B 为 NULL，则返回 NULL；如果 A 小于或等于 B，则返回 TRUE，否则返回 FALSE
A>B	基本数据类型	如果 A 或者 B 为 NULL，则返回 NULL；如果 A 大于 B，则返回 TRUE，否则返回 FALSE
A>=B	基本数据类型	如果 A 或者 B 为 NULL，则返回 NULL；如果 A 大于或等于 B，则返回 TRUE，否则返回 FALSE
A [NOT] BETWEEN B AND C	基本数据类型	如果 A，B 或者 C 任一为 NULL，则结果为 NULL；如果 A 的值大于或等于 B，并且小于或等于 C，则结果为 TRUE，否则为 FALSE。使用 NOT 关键字可以达到相反的结果
A IS NULL	所有数据类型	如果 A 等于 NULL，则返回 TRUE，否则返回 FALSE
A IS NOT NULL	所有数据类型	如果 A 不等于 NULL，则返回 TRUE，否则返回 FALSE
A [NOT] LIKE B	STRING 类型	B 是一个 SQL 下的简单正则表达式，如果 A 与其匹配，则返回 TRUE，否则返回 FALSE。B 的表达式说明如下：'x%'表示 A 必须是以字母'x'开头；'%x'表示 A 必须以字母'x'结尾；'%x%'表示 A 包含有字母'x'，可以位于开头、结尾或者字符串中间。下划线'_'匹配单个字符。B 必须要和整个字符串 A 相匹配才行。使用关键字 NOT 可以达到相反的效果
A RLIKE B, A REGEXP B	STRING 类型	B 是一个正则表达式，如果 A 与其匹配，则返回 TRUE，否则返回 FALSE。匹配使用的是 JDK 中的正则表达式接口实现的，因为正则表达式也依据它的规则。例如，正则表达式必须和整个字符串 A 相匹配，而不是只需与其子字符串匹配

2. LIKE 和 RLIKE

LIKE 和 RLIKE 谓词操作符的实例。

```
hive> SELECT * FROM sougou_20111230 WHERE keyword LIKE '%videos';
OK
20111230000018  596444b8c02b7b30c11273d5bbb88741    pissing videos 1    1
http://lakery.com/8y18-girl-pissing-in-mouth?nearest
20111230000027  9c89762b968568aaa0bed63579088f8e    stocking videos 3   1
http://www.sexystockingtops.com/updates/update.html
Time taken: 0.296 seconds, Fetched: 2 row(s)
```

RLIKE 子句是 LIKE 这个功能的一个扩展,也是通过 Java 的正则表达式来指定匹配条件。RLIKE 的用法实例。

```
hive> SELECT * FROM sougou_20111230
    > WHERE keyword RLIKE
    > '.*(in|ing).*';
OK
20111230000018  596444b8c02b7b30c11273d5bbb88741    pissing videos 1    1
http://lakery.com/8y18-girl-pissing-in-mouth?nearest
20111230000019  11e2e89dbf484ed187e73cbeaf1e0084    www.june9.info@16   1
1         http://r.baidu.com/1QCa0
20111230000019  63fd6f826a5f83d795f08778468d0e14    yunvxinjin      4   1
http://www.zvod.net/zvoddianying/14028.html
20111230000027  9c89762b968568aaa0bed63579088f8e    stocking videos 3   1
http://www.sexystockingtops.com/updates/update.html
Time taken: 0.209 seconds, Fetched: 4 row(s)
```

关键字 RLIKE 后面的字符串表达的含义。

(1) 点(.)表示和任一的字符匹配。

(2) 星号(*)表示重复"左边的字符串"零次到无数次。

(3) 表达式(x|y)表示和 x 或者 y 匹配。

3.2.3 GROUP BY 语句与 HAVING 语句

GROUP BY 子句按照一个或者多个列对结果进行分组,并对每个组执行聚合操作。GROUP BY 子句通常会和聚合函数一起使用(聚合函数在 5.1.4 小节会介绍到)。

GROUP BY 语句实例。

```
hive>SELECT keyword,count(keyword)FROM sougou_20111230
    >GROUP BY keyword;
```

结果如下:

```
12306.cn         1
baidu    2
clearaudio basic         1
fdf      1
pissing  videos 1
stocking videos 1
www.june9.info@16        1
youku    1
yunvxinjin       1
```

HAVING 子句允许用户通过一个简单的语法,完成原本需要通过子查询才能对 GROUP BY 语句产生的分组进行条件过滤的任务。例如:

```
hive>SELECT rank,searchorder,count(*)FROM sougou_20111230
    >GROUP BY rank,searchorder
    >HAVING rank>3;
```

结果如下:

注意：GROUP BY 语句后面跟条件语句只能是 HAVING 条件语句，而不能是 WHERE 条件语句。

3.2.4 JOIN 语句

Hive 支持常用的 SQL JOIN 语句，但是只支持等值连接。

1. 内连接（INNER JOIN）

内连接中，只有进行连接的两个表中都存在与连接标准相匹配的数据，才会被保留下来。接下来，就让表 sougou_20111230 和表 sougou_old_20111230 做内连接。

```
hive>SELECT a.* FROM sougou_20111230 a
    >JOIN sougou_old_20111230 b ON a.uid=b.uid
    >WHERE a.rank>1;
```

JOIN 操作的结果为：

```
20111230000019  63fd6f826a5f83d795f08778468d0e14    yunvxinjin    4    1
http://www.zvod.net/zvoddianying/14028.html
20111230000027  4a6f0d5cc0bcf16e32e74ae49663b60d    baidu  2    1    h
ttp://site.baidu.com/
20111230000027  9c89762b968568aaa0bed63579088f8e    stocking videos 3    1
http://www.sexystockingtops.com/updates/update.html
20111230000041  ec0363079f36254b12a5e30bdc070125    clearaudio basic    4
1    http://alatest.hk/reviews/receivers-amplifiers/clearaudio-balance/po3-32
225900,86/
Time taken: 37.962 seconds, Fetched: 4 row(s)
```

2. 外连接（OUTER JOIN）

外连接有左外连接（LEFT OUTER JOIN）、右外连接（RIGHT OUTER JOIN）和完全外连接（FULL OUTER JOIN）3 种。

（1）左外连接

在这种 JOIN 操作中，JOIN 操作符左边表中符合条件的所有记录将会被返回；JOIN 操作符右边表中，如果没有符合 ON 的判定条件时，从右边表指定选择的列的值将会是 NULL。例如：

```
hive>SELECT a.uid,a.keyword,b.uid,b.keyword
    >FROM sougou_old_20111230 a
    >LEFT OUTER JOIN sougou_20111230 b ON a.uid=b.uid;
```

JOIN 操作的结果为：

```
698956eb07815439fe5f46e9a4503997      youku   698956eb07815439fe5f46e9a4503997      youku
f31f594bd1f3147298bd952ba35de84d      12306.cn      f31f594bd1f3147298bd952ba35de84d      12306.cn
596444b8c02b7b30c11273d5bbb88741      pissing videos 596444b8c02b7b30c11273d5bbb88741      pissing videos
11e2e89dbf484ed187e73cbeaf1e0084      www.june9.info@16      11e2e89dbf484ed187e73cbeaf1e0084      www.june9.i
nfo@16
63fd6f826a5f83d795f08778468d0e14      yunvxinjin      63fd6f826a5f83d795f08778468d0e14      yunvxinjin
4a6f0d5cc0bcf16e32e74ae49663b60d      baidu   4a6f0d5cc0bcf16e32e74ae49663b60d      baidu
9c89762b968568aaa0bed63579088f8e      stocking videos 9c89762b968568aaa0bed63579088f8e      stocking videos
4c4c23ff94387248f4dc88166177058a      baidu   4c4c23ff94387248f4dc88166177058a      baidu
66c5bb7774e31d0a22278249b26bc83a      fanrenxiuxianzhuan      NULL    NULL
b97920521c78de70ac38e3713f524b50      benbenlianmeng  NULL    NULL
f2f5a21c764aebde1e8afcc2871e086f      zaixiandaili    NULL    NULL
96994a0480e7e1edcaef67b20d8816b7      weidadaoyan     NULL    NULL
698956eb07815439fe5f46e9a4503997      youku   698956eb07815439fe5f46e9a4503997      youku
ec0363079f36254b12a5e30bdc070125      clearaudio basic        ec0363079f36254b12a5e30bdc070125      clearaudio
basic
6da1dcbaeab299deffe5932d902e775d      lme     NULL    NULL
```

(2) 右外连接

在这种 JOIN 操作中，JOIN 操作符右边表中符合条件的所有记录将会被返回；JOIN 操作符左边表中，如果没有符合 ON 的判定条件时，那么从左边表指定选择的列的值将会是 NULL。例如：

```
hive>SELECT a.uid,a.keyword,b.uid,b.keyword
    >FROM sougou_old_20111230 a
    >RIGHT OUTER JOIN sougou_20111230 b ON a.uid=b.uid;
```

JOIN 操作的结果为：

```
698956eb07815439fe5f46e9a4503997        youku   698956eb07815439fe5f46e9a4503997        youku
698956eb07815439fe5f46e9a4503997        youku   698956eb07815439fe5f46e9a4503997        youku
f31f594bd1f3147298bd952ba35de84d        12306.cn        f31f594bd1f3147298bd952ba35de84d        12306.cn
596444b8c02b7b30c11273d5bbb88741        pissing videos  596444b8c02b7b30c11273d5bbb88741        pissing videos
11e2e89dbf484ed187e73cbeaf1e0084        www.june9.info@16       11e2e89dbf484ed187e73cbeaf1e0084        www.june9.i
nfo@16
63fd6f826a5f83d795f08778468d0e14        yunvxinjin      63fd6f826a5f83d795f08778468d0e14        yunvxinjin
NULL    NULL    637b29b47fed3853e117aa7009a4b621        fdf
4a6f0d5cc0bcf16e32e74ae49663b60d        baidu   4a6f0d5cc0bcf16e32e74ae49663b60d        baidu
9c89762b968568aaa0bed63579088f8e        stocking videos 9c89762b968568aaa0bed63579088f8e        stocking videos
ec0363079f36254b12a5e30bdc070125        clearaudio basic        ec0363079f36254b12a5e30bdc070125        clearaudio
basic
4c4c23ff94387248f4dc88166177058a        baidu   4c4c23ff94387248f4dc88166177058a        baidu
```

(3) 完全外连接

在进行完全外连接的 JOIN 操作时，将会返回所有表中符合条件的所有记录；如果任一张表的指定字段没有符合条件的值，那么就返回 NULL。例如：

```
hive>SELECT a.uid,a.keyword,b.uid,b.keyword
    >FROM sougou_old_20111230 a
    >FULL OUTER JOIN sougou_20111230 b ON a.uid=b.uid;
```

JOIN 操作的结果显示为：

```
11e2e89dbf484ed187e73cbeaf1e0084        www.june9.info@16       11e2e89dbf484ed187e73cbeaf1e0084        www.june9.i
nfo@16
4a6f0d5cc0bcf16e32e74ae49663b60d        baidu   4a6f0d5cc0bcf16e32e74ae49663b60d        baidu
4c4c23ff94387248f4dc88166177058a        baidu   4c4c23ff94387248f4dc88166177058a        baidu
596444b8c02b7b30c11273d5bbb88741        pissing videos  596444b8c02b7b30c11273d5bbb88741        pissing videos
NULL    NULL    637b29b47fed3853e117aa7009a4b621
63fd6f826a5f83d795f08778468d0e14        yunvxinjin      63fd6f826a5f83d795f08778468d0e14        yunvxinjin
66c5bb7774e31d0a22278249b26bc83a        fanrenxiuxianzhuan      NULL    NULL
698956eb07815439fe5f46e9a4503997        youku   698956eb07815439fe5f46e9a4503997        youku
698956eb07815439fe5f46e9a4503997        youku   698956eb07815439fe5f46e9a4503997        youku
6da1dcbaeab299deffe5932d902e775d        lme     NULL    NULL
96994a0480e7e1edcaef67b20d8816b7        weidadaoyan     NULL    NULL
9c89762b968568aaa0bed63579088f8e        stocking videos 9c89762b968568aaa0bed63579088f8e        stocking videos
b97920521c78de70ac38e3713f524b50        benbenlianmeng  NULL    NULL
ec0363079f36254b12a5e30bdc070125        clearaudio basic        ec0363079f36254b12a5e30bdc070125        clearaudio
basic
f2f5a21c764aebde1e8afcc2871e086f        zaixiandaili    NULL    NULL
f31f594bd1f3147298bd952ba35de84d        12306.cn        f31f594bd1f3147298bd952ba35de84d        12306.cn
```

3. 左半开连接（LEFT SEMI JOIN）

对于常见的内连接来说，左半开连接是特殊的、已优化的。左半开连接是通过关键字 LEFT SEMI JOIN 进行标识的。在 JOIN 操作时，它会返回左边表的记录，但是这些左边表的记录对于右边表满足 ON 语句中的判定条件。例如：

```
hive>SELECT * FROM sougou_old_20111230 a
    >LEFT SEMI JOIN sougou_20111230 b ON a.rank=b.rank;
```

JOIN 操作的结果显示为：

```
20111230000009  698956eb07815439fe5f46e9a4503997           youku  1        1   http://www.youku.com/
20111230000014  f31f594bd1f3147298bd952ba35de84d           12306.cn  1     1   http://www.12306.cn/
20111230000018  596444b8c02b7b30c11273d5bbb88741           pissing  videos 1  1  http://lakery.com/8y18-girl
                -pissing-in-mouth?nearest
20111230000019  11e2e89dbf484ed187e73cbeaf1e0084           www.june9.info@16       1    http://r.baidu.com/
                lQCa0
20111230000019  63fd6f826a5f83d795f08778468d0e14           yunvxinjin  4    1   http://www.zvod.net/zvoddia
                nying/14028.html
20111230000027  4a6f0d5cc0bcf16e32e74ae49663b60d           baidu  2        http://site.baidu.com/
20111230000027  9c89762b968568aaa0bed63579088f8e           stocking videos 3 1  http://www.sexystockingtops
                .com/updates/update.html
20111230000058  4c4c23ff94387248f4dc0010G177050a           baidu  1        http://www.baidu.com/
20111230000005  66c5bb7774e31d0a22278249b26bc83a           fanrenxiuxianzhuan  3  1   http://www.booksky.
                org/BookDetail.aspx?BookID=1050804&Level=1
20111230000007  b97920521c78de70ac38e3713f524b50           benbenlianmeng  1    http://www.bblianmeng.com/
20111230000008  f2f5a21c764aebde1e8afcc2871e086f           zaixiandaili  2   1   http://proxyie.cn/
20111230000009  96994a0480e7e1edcaef67b20d8816b7           weidadaoyan  1   1   http://movie.douban.com/rev
                iew/1128960/
20111230000009  698956eb07815439fe5f46e9a4503997           youku  1        http://www.youku.com/
20111230000041  ec0363079f36254b12a5e30bdc070125           clearaudio basic  4   1   http://alatest.hk/r
                eviews/receivers-amplifiers/clearaudio-balance/po3-32225900,86/
20111230000041  6da1dcbaeab299deffe5932d902e775d           lme  2     1   http://finance.sina.com.cn/money/fu
                ture/CAD/quote.shtml
```

4. 笛卡儿积 JOIN

笛卡儿积是一种连接，左边表的行数乘以右边表的行数等于笛卡儿积的大小。

在执行 JOIN 操作前，分别来看看作为实例的两张表中有多少条数据。表 sougou_old_20111230 中有 15 条数据，表 sougou_20111230 中有 10 条数据。

```
hive>SELECT * FROM sougou_old_20111230 JOIN sougou_20111230;
```

查看 JOIN 后的条数。

```
Time taken: 42.954 seconds, Fetched: 150 row(s)
```

一般情况下，很少用到笛卡儿积，但在某些情况下这种连接很有用。例如，有一个表表示用户的饮食喜好，另外一张表表示某些小吃，要求用算法来推测用户可能会喜欢的小吃时，使用笛卡儿积就能生成所有用户和所有小吃的对应关系的集合。

5. map 端连接(map-side JOIN)

如果所有表中只有一张表是小表，那么可以在最大的表通过 mapper 时将小表完全放到内存中。Hive 可以在 map 端执行连接过程，称为 map 端连接，这是因为 Hive 可以和内存中的小表逐一进行匹配，从而省略掉常规连接操作所需要的 reduce 过程，这样就节省了时间。

由于这种连接不仅减少了 reduce 过程，而且有时可以同时减少 map 过程的执行步骤，即使对于很小的数据集，也明显要快于常规的连接操作。

一般的 JOIN 操作实例。

```
hive>SELECT a.uid,a.keyword,b.uid,b.keyword
    >FROM sougou_old_20111230 a
    >JOIN sougou_20111230 b ON a.uid=b.uid;
```

所需要花费的时间。

```
Time taken: 44.605 seconds, Fetched: 10 row(s)
```

同样数据的 map-side JOIN 操作实例。

```
hive>SELECT /*+MAPJOIN(n)*/a.uid,a.keyword,b.uid,b.keyword
    >FROM sougou_old_20111230 a
    >JOIN sougou_20111230 b ON a.uid=b.uid;
```

操作的时间。

```
Time taken: 42.479 seconds, Fetched: 10 row(s)
```

显然，优化了的 map-side JOIN 所需要的时间要短。

6. JOIN 优化

下列 JOIN 优化会使 JOIN 操作节省时间和资源。

（1）当 Hive 执行 JOIN 发生内存溢出时，可以修改 Hive 的配置文件 hive-site.xml 来增大内存，如 mapred.child.java.opts -Xmx 1024m。

（2）执行 JOIN 操作时，尽量把小表放前面。大表放前面，内存溢出而出错的可能性更大，因为在对每行记录进行连接操作时，Hive 会尝试将其他表缓存起来，然后扫描最后那个表进行计算。所以用户在执行 JOIN 操作时，让连续查询中的表的大小从左到右是递增的。

（3）在连接多个表时，尽量使用相同的连接键连接 ON 子句，因为这样只会产生一个 MapReduce job。在本小节的多个例子中，每个 ON 子句中都使用了 a.uid 作为其中一个 JOIN 连接键。在这种情况下，Hive 通过一个优化可以在同一个 MapReduce job 中连接多张表。

（4）对分区表进行操作时，需要对分区进行过滤。在 JOIN 操作时，分区过滤需要放到 ON 语句或子查询中，不能放到 ON 后面的 WHERE 里。否则，会扫描所有表，最后才判断分区。也就是说，程序会先执行 JOIN 操作，然后才会执行最后的 WHERE 操作，这样就会花费大量的时间。

（5）当一个大表和一个很小的表进行 JOIN 操作时，使用 map-side JOIN。

3.2.5 ORDER BY 语句和 SORT BY 语句

ORDER BY 会对输入做全局排序，只有一个 Reducer，若数据的规模比较大，需要较长的计算时间。Hive 中的 ORDER BY 也是对一个结果集进行排序，不同于关系型数据库的是底层架构。Hive 的 hive-site.xml 配置文件中的参数 hive.mapred.mode 控制着 Hive 的执行方式：若选择 strict，则 ORDER BY 需要指定 LIMIT，若有分区，则需要指定哪个分区；若为 nostrict，则与关系型数据库差不多。例如：

```
hive>SELECT a.logdate,a.uid,a.keyword
    >FROM sougou_20111230 a
    >ORDER BY a.logdate ASC;
```

排序结果为：

```
20111230000009    698956eb07815439fe5f46e9a4503997    youku
20111230000014    f31f594bd1f3147298bd952ba35de84d    12306.cn
20111230000018    596444b8c02b7b30c11273d5bbb88741    pissing videos
20111230000019    63fd6f826a5f83d795f08778468d0e14    yunvxinjin
20111230000019    11e2e89dbf484ed187e73cbeaf1e0084    www.june9.info@16
20111230000020    637b29b47fed3853e117aa7009a4b621    fdf
20111230000027    9c89762b968568aaa0bed63579088f8e    stocking videos
20111230000027    4a6f0d5cc0bcf16e32e74ae49663b60d    baidu
20111230000041    ec0363079f36254b12a5e30bdc070125    clearaudio basic
20111230000058    4c4c23ff94387248f4dc88166177058a    baidu
```

Hive 还增加了一个可供选择的方式——SORT BY,其操作只会在每个 reducer 中对数据进行排序,也就是执行一个局部排序的过程。这可以保证每个 reducer 的输出数据都是有序的,但是并非全局有序。

sort 可以控制每个 reducer 产生的文件都是排序,再对多个排序好的文件做二次归并排序。SORT BY 特点如下:

(1) SORT BY 受 hive.mapred.mode 是 strict 或 nonstrict 的影响,但若有分区需要指定分区。

(2) SORT BY 的数据在同一个 reducer 中数据是按指定字段排序。

(3) SORT BY 可以指定执行的 reducer 个数,如 set mapred.reducer.tasks=5,对输出的数据再执行归并排序,即可以得到全部结果。

SORT BY 举例。

```
hive>SELECT a.logdate,a.keyword,a.url
    >FROM sougou_20111230 a
    >SORT BY a.logdate DESC;
```

排序结果为:

```
20111230000058    baidu       http://www.baidu.com/
20111230000041    clearaudio basic       http://alatest.hk/reviews/receivers-amplifiers/cle
900,86/
20111230000027    stocking videos http://www.sexystockingtops.com/updates/update.html
20111230000027    baidu       http://site.baidu.com/
20111230000020    fdf         http://www.163pan.com/files/70z00000j.html
20111230000019    yunvxinjin      http://www.zvod.net/zvoddianying/14028.html
20111230000019    www.june9.info@16       http://r.baidu.com/1QCa0
20111230000018    pissing     videos http://lakery.com/8y18-girl-pissing-in-mouth?nearest
20111230000014    12306.cn        http://www.12306.cn/
20111230000009    youku       http://www.youku.com/
```

对于这两种情况,语法上的区别在于一种的关键字是 ORDER;另一种的关键字是 SORT。用户可以指定任意期望进行排序的字段,并且在后面加上是升序或者降序的关键字即可。升序的关键字是 ASC(默认的),降序的关键字是 DESC。

3.2.6 CLUSTER BY 语句

CLUSTER BY 语句的排序只能是倒序排序,不能指定排序 ASC 或者 DESC,例如:

```
hive>SELECT a.logdate,a.uid,a.keyword
    >FROM sougou_20111230 a
    >CLUSTER BY a.keyword;
```

排序结果为:

```
20111230000014    f31f594bd1f3147298bd952ba35de84d        12306.cn
20111230000058    4c4c23ff94387248f4dc88166177058a        baidu
20111230000027    4a6f0d5cc0bcf16e32e74ae49663b60d        baidu
20111230000041    ec0363079f36254b12a5e30bdc070125        clearaudio basic
20111230000020    637b29b47fed3853e117aa7009a4b621        fdf
20111230000018    596444b8c02b7b30c11273d5bbb88741        pissing  videos
20111230000027    9c89762b968568aaa0bed63579088f8e        stocking videos
20111230000019    11e2e89dbf484ed187e73cbeaf1e0084        www.june9.info@16
20111230000009    698956eb07815439fe5f46e9a4503997        youku
20111230000019    63fd6f826a5f83d795f08778468d0e14        yunvxinjin
```

3.2.7 UNION ALL 语句

Hive 不支持 UNION 操作,只支持 UNION ALL。UNION ALL 可以将两个或多个表进行合并。每一个子查询都必须具有相同的列,并且对应的每个字段类型也必须一致。例如:

```
hive>SELECT c.uid,c.keyword
    >FROM(SELECT a.uid,a.keyword FROM sougou_20111230 a
    >UNION ALL
    >SELECT b.uid,b.keyword FROM sougou_old_20111230 b)c;
```

合并结果为:

```
698956eb07815439fe5f46e9a4503997        youku
f31f594bd1f3147298bd952ba35de84d        12306.cn
596444b8c02b7b30c11273d5bbb88741        pissing videos
11e2e89dbf484ed187e73cbeaf1e0084        www.june9.info@16
63fd6f826a5f83d795f08778468d0e14        yunvxinjin
4a6f0d5cc0bcf16e32e74ae49663b60d        baidu
9c89762b968568aaa0bed63579088f8e        stocking videos
4c4c23ff94387248f4dc88166177058a        baidu
66c5bb7774e31d0a22278249b26bc83a        fanrenxiuxianzhuan
b97920521c78de70ac38e3713f524b50        benbenlianmeng
f2f5a21c764aebde1e8afcc2871e086f        zaixiandaili
96994a0480e7e1edcaef67b20d8816b7        weidadaoyan
698956eb07815439fe5f46e9a4503997        youku
ec0363079f36254b12a5e30bdc070125        clearaudio basic
6da1dcbaeab299deffe5932d902e775d        lme
698956eb07815439fe5f46e9a4503997        youku
f31f594bd1f3147298bd952ba35de84d        12306.cn
596444b8c02b7b30c11273d5bbb88741        pissing videos
11e2e89dbf484ed187e73cbeaf1e0084        www.june9.info@16
63fd6f826a5f83d795f08778468d0e14        yunvxinjin
637b29b47fed3853e117aa7009a4b621        fdf
4a6f0d5cc0bcf16e32e74ae49663b60d        baidu
9c89762b968568aaa0bed63579088f8e        stocking videos
ec0363079f36254b12a5e30bdc070125        clearaudio basic
4c4c23ff94387248f4dc88166177058a        baidu
```

对于 UNION ALL 语句,需要注意的是:

(1) 子查询相当于表名,使用 FROM 关键字需要指定真实表名或表别名。

(2) Hive 不支持 UNION,只支持 UNION ALL。

(3) 子查询中使用 UNION ALL 时,在子查询里不能使用 count、sum 等聚合函数,这类函数在 5.1.4 小节会详细介绍。

(4) 两表直接进行 UNION ALL 可以使用 count、sum 等聚合函数。

(5) 两张表进行 UNION ALL 取相同的字段名称,可正常输出指定数据内容,且结果为两张表的结果集。

3.3 抽样查询

对于非常大的数据集,有时用户需要使用的是一个具有代表性的查询语句,而不是全部结果。在 Hive 中,可以通过分桶抽样来满足这样的需求。分桶语句中的分母表示的是数据将会被散列的桶的个数,而分子表示的是将会选择的桶的个数。

表 sougou_20111230 中有 10 条数据，可以使用数学函数 rand() 进行抽样，这个函数会从 10 条数据中随机返回 n 条数据。

```
hive> SELECT * FROM sougou_20111230 TABLESAMPLE(BUCKET 3 OUT OF 10 ON rand())s;
OK
20111230000014  f31f594bd1f3147298bd952ba35de84d       12306.cn         1          http://www.12306.cn/
Time taken: 0.104 seconds, Fetched: 1 row(s)
hive> SELECT * FROM sougou_20111230 TABLESAMPLE(BUCKET 3 OUT OF 10 ON rand())s;
OK
20111230000041  ec0363079f36254b12a5e30bdc070125       clearaudio basic    4     1   http://alatest
.hk/reviews/receivers-amplifiers/clearaudio-balance/po3-32225900,86/
Time taken: 0.091 seconds, Fetched: 1 row(s)
hive> SELECT * FROM sougou_20111230 TABLESAMPLE(BUCKET 3 OUT OF 10 ON rand())s;
OK
20111230000009  698956eb07815439fe5f46e9a4503997       youku   1     1   http://www.youku.com/
20111230000058  4c4c23ff94387248f4dc88166177058a       baidu   1     1   http://www.baidu.com/
Time taken: 0.063 seconds, Fetched: 2 row(s)
```

rand() 数学函数在表 5-1 中有介绍，它就是数学函数中的一个随机函数。不仅可以用 rand() 函数，还可以用指定的列。这时同一语句执行多次返回的结果是相同的。

```
hive> SELECT * FROM sougou_20111230 TABLESAMPLE(BUCKET 3 OUT OF 4 ON rank)s;
OK
20111230000027  4a6f0d5cc0bcf16e32e74ae49663b60d       baidu   2     1   http://site.baidu.com/
Time taken: 0.067 seconds, Fetched: 1 row(s)
hive> SELECT * FROM sougou_20111230 TABLESAMPLE(BUCKET 3 OUT OF 4 ON rank)s;
OK
20111230000027  4a6f0d5cc0bcf16e32e74ae49663b60d       baidu   2     1   http://site.baidu.com/
Time taken: 0.082 seconds, Fetched: 1 row(s)
hive> SELECT * FROM sougou_20111230 TABLESAMPLE(BUCKET 2 OUT OF 4 ON rank)s;
OK
20111230000009  698956eb07815439fe5f46e9a4503997       youku    1     1    http://www.youku.com/
20111230000014  f31f594bd1f3147298bd952ba35de84d       12306.cn 1     1    http://www.12306.cn/
20111230000018  596444b8c02b7b30c11273d5bbb88741       pissing videos 1  1 http://lakery.com/8y18
-girl-pissing-in-mouth?nearest
20111230000019  11e2e89dbf484ed187e73cbeaf1e0084       www.june9.info@16   1   1   http://r.baidu
.com/1QCa0
20111230000020  637b29b47fed3853e117aa7009a4b621       fdf      1     1    http://www.163pan.com/files/70
z00000j.html
20111230000058  4c4c23ff94387248f4dc88166177058a       baidu    1     1    http://www.baidu.com/
Time taken: 0.085 seconds, Fetched: 6 row(s)
```

3.3.1 数据块抽样

在 Hive 中，也可以按照抽样百分比进行抽样，这种抽样方式是基于行数的，按照输入路径下的数据块百分比进行抽样。例如：

```
hive> SELECT * FROM sougou_20111230 TABLESAMPLE(1.0 PERCENT)s;
OK
20111230000009  698956eb07815439fe5f46e9a4503997       youku   1     1   http://www.youku.com/
Time taken: 0.057 seconds, Fetched: 1 row(s)
```

注意：这种抽样方式不一定适用于所有的文件格式。这种抽样的最小抽样单元是一个 HDFS 数据块。因此，如果表的数据块大小小于普通的块大小 128MB，那么将会返回所有行。

3.3.2 分桶表的输入裁剪

TABLESAMPLE 还可以用另外一种方式表示，例如：

```
hive> SELECT * FROM sougou_20111230 WHERE rank%2=0;
OK
20111230000019  63fd6f826a5f83d795f08778468d0e14       yunvxinjin   4   1   http://www.zvod.net/zv
oddianying/14028.html
20111230000027  4a6f0d5cc0bcf16e32e74ae49663b60d       baidu   2     1   http://site.baidu.com/
20111230000041  ec0363079f36254b12a5e30bdc070125       clearaudio basic   4   1   http://alatest
.hk/reviews/receivers-amplifiers/clearaudio-balance/po3-32225900,86/
Time taken: 0.1 seconds, Fetched: 3 row(s)
```

抽样会扫描表中的所有数据,然后在每 N 行中抽取一行数据。对于大多数的表是这样的,但是如果 TABLESAMPLE 语句中指定的列和 CLUSTERED BY 语句中指定的列相同,那么 TABLESAMPLE 查询就只会扫描涉及的表的 hash 分区下的数据。

例如,先创建一个这样的桶表。

```
hive>CREATE TABLE sougou_bucketed(
    >logdate STRING,
    >uid STRING,
    >keyword STRING,
    >rank INT,
    >url STRING)
    >CLUSTERED BY(rank)
    >INTO 3 BUCKETS
    >ROW FORMAT DELIMITED
    >FIELDS TERMINATED BY '\t'
    >LOCATION '/data/sougou_bucketed';
```

然后,设置 Hive 中的分桶配置,并且把原来的数据插入这个桶表中。

```
hive>SET hive.enforce.bucketing=true;
hive>INSERT OVERWRITE TABLE sougou_bucketed
    >SELECT * FROM sougou_20111230;
```

查看表路径下分的数据桶信息。

```
hive> dfs -ls /data/sougou_bucketed;
Found 3 items
-rw-r--r--   1 zkpk supergroup        120 2016-03-30 20:55 /data/sougou_bucketed/000000_0
-rw-r--r--   1 zkpk supergroup        816 2016-03-30 20:55 /data/sougou_bucketed/000001_0
-rw-r--r--   1 zkpk supergroup         81 2016-03-30 20:55 /data/sougou_bucketed/000002_0
```

查看其中一个表中的 1 个数据桶。

```
hive>dfs -cat /data/sougou_bucketed/000000_0;
hive> dfs -cat /data/sougou_bucketed/000000_0;
20111230000027  9c89762b968568aaa0bed63579088f8e        stocking videos 3       1       http://www.sexystockin
gtops.com/updates/update.html
```

此时,表已经聚集成了 3 个数据桶,下面的查询可以仅对其中一个数据桶进行高效抽样。

```
hive> SELECT * FROM sougou_bucketed TABLESAMPLE(BUCKET 2 OUT OF 3 ON rank)s;
OK
20111230000058  4c4c23ff94387248f4dc88166177058a        baidu   1       1       http://www.baidu.com/
20111230000041  ec0363079f36254b12a5e30bdc070125        clearaudio basic        4       1       http://alatest
.hk/reviews/receivers-amplifiers/clearaudio-balance/po3-32225900,86/
20111230000020  637b29b47fed3853e117aa7009a4b621        fdf     1       1       http://www.163pan.com/files/70
z00000j.html
20111230000019  63fd6f826a5f83d795f08778468d0e14        yunvxinjin      4       1       http://www.zvod.net/zv
oddianying/14028.html
20111230000019  11e2e89dbf484ed187e73cbeaf1e0084        www.june9.info@16       1       1       http://r.baidu
.com/1QCa0
20111230000018  596444b8c02b7b30c11273d5bbb88741        pissing videos 1        1       http://lakery.com/8y18
-girl-pissing-in-mouth?nearest
20111230000014  f31f594bd1f3147298bd952ba35de84d        12306.cn        1       1       http://www.12306.cn/
20111230000009  698956eb07815439fe5f46e9a4503997        youku   1       1       http://www.youku.com/
Time taken: 0.079 seconds, Fetched: 8 row(s)
```

本章小结

（1）掌握 3 种数据加载的方法、动态分区插入数据，以及 3 种数据导出的方法。

（2）掌握 SELECT…FROM 查询语句，以及包含在 SELECT…FROM 查询语句中的其他语句。

（3）掌握 WHERE 语句、GROUP BY 语句、JOIN 语句，要重点掌握 JOIN 优化。

（4）掌握 ORDER BY 语句、SORT BY 语句、CLUSTER BY 语句和 UNION ALL 语句。

（5）了解抽样查询中的数据块抽样和分桶表的输入裁剪。

习 题

1. 选择题

（1）Hive 表中的数据可以导出到（　　）。
　　A. 本地文件系统　　　　　　B. HDFS 文件系统
　　C. 另一张表　　　　　　　　D. 以上都可以

（2）在 group by 语句后面可以跟的条件语句是（　　）。
　　A. having　　B. where　　C. if　　D. while

（3）在 JOIN 连接中，（　　）连接不存在。
　　A. LEFT OUTER JOIN　　　　B. RIGHT OUTER JOIN
　　C. LEFT SEMI JOIN　　　　　D. RIGHT SEMI JOIN

（4）以下语句中，不能指定排序 ASC 或者 DESC 的是（　　）。
　　A. ORDER BY　　B. SORT BY　　C. CLUSTER BY　　D. 以上都不能

（5）下列选项中，对于 UNION ALL 的描述正确的是（　　）。
　　A. Hive 支持 UNION，不支持 UNION ALL
　　B. 两表直接进行 UNION ALL 可以使用 count、sum 等聚合函数
　　C. 子查询中使用 UNION ALL 时，在子查询里可以使用 count、sum 等聚合函数
　　D. 两张表进行 UNION ALL 取相同的字段名称，不能正常输出指定数据内容，且结果为两张表的结果集

2. 问答题

（1）本章介绍了几种数据加载方法？分别是什么？它们的语法分别是什么？

（2）简述动态分区插入和它的应用场景。

（3）ORDER BY 语句和 SORT BY 语句的区别是什么？

（4）简述 JOIN 优化。

（5）简述运用 UNION ALL 语句时需要注意的是什么？

（6）什么是抽样查询？什么是数据块抽样？

第4章 HiveQL 视图和索引

本章摘要

HiveQL 中的视图和索引与传统数据库中的视图和索引有许多相似之处。本章对视图和索引的学习都是以实例的形式展开进行学习。首先，简单地了解视图，接着学习怎样创建视图、怎样显示视图和删除视图；索引也从这几个方面类似地展开学习。

4.1 视图

视图是虚拟表，其内容由查询来定义。每个视图只允许保存一个查询，并且像对待表一样对这个查询进行操作。这是一个逻辑结构，Hive 目前不支持物化视图，因为它不像表一样会存储数据。

视图是存储在数据库中的查询的 SQL 语句，它的存在主要有两种原因：一种是安全原因，视图可以隐藏一些数据；另一种是可使复杂的查询易于理解和使用。

从用户角度来看，每个视图是从一个特定的角度来查看数据库中的数据的。从数据库内部来看，视图是由 SELECT 语句组成的查询定义的虚拟表，也可以说，视图是由一张或多张表中的数据组成的。从数据库外部来看，每个视图就如同一张表一样，对表能够进行的一般操作都可以应用于视图中。

同真实的表一样，视图的作用类似于筛选。定义视图的筛选可以来自当前或其他数据库的一个或多个表，或者其他视图。

当一个查询引用一个视图的时候，这个视图所定义的查询语句将和用户的查询语句组合在一起，然后供 Hive 制订查询计划。从逻辑上讲，可以想象为 Hive 先执行这个视图，然后使用这个结果进行后续的查询。

4.1.1 创建视图

创建视图，通俗来讲，就是把用户需要的数据从一个大表中拿出来放在一个小的表中，具有更高的查询效果。

【例 4-1】 给表 sougou_20111230 创建一个视图。

```
hive>CREATE VIEW sougou_total_view_20111230
    >AS
    >SELECT * FROM sougou_20111230;
```

例 4-1 是以表中的 6 个字段的数据创建的一个视图，往往并不需要表中所有的字段数据，而只用部分字段。例如，用查询表 sougou_20111230 中的 3 个字段来创建一个视图。

```
hive>CREATE VIEW sougou_view_20111230(logdate,uid,keyword)
    >AS
    >SELECT logdate,uid,keyword FROM sougou_20111230;
```

当数据库中创建的视图较多时，为了避免重复，在创建视图的时候可以在关键字 VIEW 后面加上 IF NOT EXISTS 子句，表示创建的视图存在就不执行后面的操作。

不仅数据的字段可以选择，记录也可以用 WHERE 子句进行筛选。不用 WHERE 子句，默认是查询表中符合条件的所有的数据，用 WHERE 子句可以更精准地过滤符合条件的数据，例如：

```
hive>CREATE VIEW sougou_rank_view_20111230(
    >logdate,uid,keyword,rank)
    >AS
    >SELECT logdate,uid,keyword,rank FROM sougou_20111230
    >WHERE rank>1;
```

4.1.2 显示视图

显示视图分为显示视图的表结构、视图的数据记录两个功能。

1. 显示视图的表结构

显示视图的表结构，需要使用关键字 DESCRIBE（有时简写成 DESC）。例如，分别来看 4.1.1 小节创建的 3 个视图。

【例 4-2】 查看视图 sougou_total_view_20111230 的表结构。

```
hive>DESCRIBE sougou_total_view_20111230;
OK
date                    string
uid                     string
keyword                 string
rank                    int
order                   int
url                     string
Time taken: 0.168 seconds, Fetched: 6 row(s)
```

【例 4-3】 查看视图 sougou_view_20111230 的表结构。

```
hive>DESC sougou_view_20111230;
OK
date                    string
uid                     string
keyword                 string
Time taken: 0.301 seconds, Fetched: 3 row(s)
```

【例 4-4】 查看视图 sougou_rank_view_20111230 的表结构。

```
hive>DESC sougou_rank_view_20111230;
OK
date                    string
uid                     string
keyword                 string
rank                    int
Time taken: 0.129 seconds, Fetched: 4 row(s)
```

2. 查看视图数据记录

创建视图就相当于新建了一张表,查看这张表中的数据就用表的查询语句。

【例 4-5】

```
Hive>SELECT * FROM sougou_rank_view_20111230;
OK
20111230000019    63fd6f826a5f83d795f08778468d0e14    yunvxinjin      4
20111230000027    4a6f0d5cc0bcf16e32e74ae49663b60d    baidu    2
20111230000027    9c89762b968568aaa0bed63579088f8e    stocking videos 3
20111230000041    ec0363079f36254b12a5e30bdc070125    clearaudio basic    4
Time taken: 0.906 seconds, Fetched: 4 row(s)
```

4.1.3 删除视图

视图往往是根据需求建立的,所以它只是暂时存在数据库中,当视图无用时可以选择删除,以节省空间。

【例 4-6】 删除视图 sougou_total_view_20111230。

```
hive> DROP VIEW sougou_total_view_20111230;
OK
Time taken: 0.914 seconds
```

注意:在删除视图的语句中可以在关键字 DROP VIEW 后面加上 IF EXISTS 子句,这样可以判断此视图是否存在,有则进行删除操作。

删除操作以后,可以查看是否删除成功。

```
hive> SHOW tables;
OK
sougou_20111230
sougou_bucketed
sougou_old_20111230
Time taken: 0.051 seconds, Fetched: 3 row(s)
```

4.2 索 引

索引是一个单独的、物理的数据库结构,它是某个表中一列或若干列值的集合和相应的逻辑指针清单(指向表中标识这些值的数据页)。通俗来讲,索引就相当于图书的目录,可以根据目录中的页码快速找到所需要的内容。当表中有大量记录时,若要对表进行查询,第一种搜索信息方式是全表搜索,将所有记录一一取出,和查询条件进行一一对比,然后返回满足条件的记录,这样做会消耗大量数据库系统时间,并且造成大量的磁盘 I/O 操作;第二种就是在表中建立索引,在索引中找到符合查询条件的索引值,然后通过保存在索引中的页码(ROWID)快速找到表中对应的记录,省时又省力。

Hive 中的索引和关系型数据库中的索引一样,维护索引也需要额外的存储空间,同时创建索引也需要消耗计算资源。因此,需要进行仔细评估才能使用索引,需要在建立索引为查询带来的好处和需要付出的代价之间做出权衡。

Hive 中的索引的功能非常有限。在关系数据库中,索引是一种与表有关的数据库结构,可以使对应于表的 SQL 语句执行得更快。Hive 中没有普通关系型数据库中键的概念,还是可以对一些字段建立索引来加速某些操作。另外,索引处理模块被设计成为可以定制的 Java 编码的插件，因此用户可以根据需要对其进行实现，以满足自身的需求。

当逻辑分区太多太细而无法使用的时候，建立索引也就成为分区的另一个不错的选择。建立索引可以帮助裁减掉一张表的一些数据块，这样能够减少 MapReduce 的输入数据量。并非所有的查询都可以通过建立索引获得好处，可以通过 EXPLAIN 命令查看某个查询语句是否用到了索引。

4.2.1 创建索引

【例 4-7】 给表 sougou_20111230 创建一个索引。

```
hive>CREATE INDEX sougou_index
    >ON TABLE sougou_20111230(uid)
    >AS 'org.apache.hadoop.hive.ql.index.compact.CompactIndexHandler'
    >WITH DEFERRED REBUILD
    >IN TABLE sougou_index_table
    >ROW FORMAT DELIMITED
    >FIELDS TERMINATED BY '\t'
    >LOCATION '/data/sougou_index_table'
    >COMMENT 'sougou_20111230 indexed by uid';
```

AS 子句指定了索引处理器，也就是一个实现了索引接口的 Java 类。Hive 本身包含了一些典型的索引实现，这里所展示的 CompactIndexHandler 就是其中的一个实现。可以通过用户的实现来优化地处理特定的场景，支持特定的文件格式等。

WITH DEFERRED REBUILD 子句在执行更新数据语句 alter index xxx_index on xxx rebuild 时，将调用 generateIndexBuildTaskList 获取 Index 的 MapReduce 并执行，为索引填充数据。

IN TABLE…子句表示索引处理器在一张新表中保留索引数据。除此之外，例子中使用到的 ROW FORMAT、FIELDS、LOCATION、COMMENT 和没有使用到的 STORED AS、STORED BY 的含义在之前我们学习过，在这里就不再提了。

在 Hive v0.8.0 版本以及之后的版本还提供了一个内置 bitmap 索引处理器。bitmap 索引普遍应用于排重后值较少的列。

【例 4-8】 对例 4-7 用 bitmap 索引处理器重写。

```
hive>CREATE INDEX sougou_index
    >ON TABLE sougou_20111230(uid)
    >AS 'BITMAP'
    >WITH DEFERRED REBUILD
    >IN TABLE sougou_index_table
    >ROW FORMAT DELIMITED
    >FIELDS TERMINATED BY '\t'
    >LOCATION '/data/sougou_index_table'
    >COMMENT 'sougou_20111230 indexed by uid';
```

4.2.2 重建索引

如果用户指定了 WITH DEFERRED REBUILD，那么新索引将呈现空白状态。在任何时候,都可以进行第一次索引创建或者使用 ALTER INDEX 对索引进行重建。例 4-7 的索引指定了 WITH DEFERRED REBUILD,对它进行了重建。

```
hive>ALTER INDEX sougou_index ON sougou_20111230 REBUILD;
```

在重建索引时,需要注意的是:

(1) 当 Hive 数据更新时,必须调用该语句更新索引。

(2) 重建索引操作是一个原子操作。因此,当重建失败时,先前构建的索引也无法使用。

注意:原子操作是指不会被线程调度机制打断的操作;这种操作一旦开始,就一直运行到结束,中间不会切换到另一个线程。

4.2.3 显示索引

【例 4-9】 查看索引。

```
hive>SHOW FORMATTED INDEX ON sougou_20111230;
OK
idx_name                tab_name              col_names             idx_tab_
name          idx_type              comment

sougou_index            sougou_20111230       uid                   sougou_i
ndex_table    compact               sougou_20111230 indexed by uid
Time taken: 0.189 seconds, Fetched: 4 row(s)
```

关键字 FORMATTED 是可选的。它的功能是使输出中包含列名称。用户还可以用 INDEXES 替换 INDEX,可以使输出中列举出多个索引信息。

4.2.4 删除索引

Hive 不允许用户使用 DROP TABLE 语句直接删除索引表,要通过删除索引来同时删除对应的索引表。

```
hive> DROP INDEX IF EXISTS sougou_index ON  sougou_20111230;
OK
Time taken: 0.428 seconds
```

来查看一下,索引是否真的被删除了。

```
hive> SHOW FORMATTED INDEX ON sougou_20111230;
OK
idx_name                tab_name              col_names             idx_tab_
name          idx_type              comment

Time taken: 0.12 seconds, Fetched: 3 row(s)
```

IF EXISTS 子句和 2.2 节学习的此子句功能一致,是为了避免抛出错误信息。如果索引的表被删除了,那么其对应的索引和索引表同时会被删除。同样地,如果原始表是一个分区表,表中的分区被删除了,那么这个分区对应的分区索引也同时会被删除。

本章小结

本章主要学习的是视图和索引,在学习完本章之后,应会对视图和索引做一些简单的操作并且会使用它。

习 题

1. 选择题

(1) 视图的内容由( )定义。
  A. 查询   B. 创建   C. 修改   D. 删除

(2) 创建了一个视图,要用( )关键字来查看视图。
  A. SHOW   B. DESCRIBE 或 DESC
  C. CREATE   D. 以上都不是

(3) 当表中有大量记录时,若要对表进行查询,有两种搜索信息的方式,( )是最优的方式。
  A. 全表搜索 B. 语句查询 C. 建立视图 D. 建立索引

(4) 在创建索引的时候,用到的关键字是( )。
  A. VIEW   B. INDEX   C. TABLE   D. 以上都不是

(5) 下列选项中,说法不正确的是( )。
  A. 视图存在主要有两种原因:一种是安全原因,视图可以隐藏一些数据;另一种是可使复杂的查询易于理解和使用
  B. 如果你的数据库中存在的视图较多,那么在创建视图时,可以在关键字 VIEW 后加上 IF NOT EXISTS 子句
  C. Hive 中的索引和关系型数据库中的索引一样,可不需要评估直接使用
  D. Hive 不允许用户直接使用 DROP TABLE 语句删除索引表

2. 问答题

(1) 简述视图的概念和带来的好处。

(2) 假如现在有一张员工表,表中包含员工的工号、员工姓名、员工职位和员工薪资。尝试创建一个视图,视图中包含员工工号和员工职位。

(3) 简述索引的概念和优缺点。

(4) 给问答题(2)中的员工表加一个索引,以员工工号为索引。

第 5 章

Hive 的函数

本章摘要

本章主要介绍 Hive 函数,包括标准函数、聚合函数、表生成函数等。为了加深对这些函数的理解,重点讲 3 个案例。通过对案例的学习,应可以独立编写用户自定义函数(UDF)、用户自定义聚合函数(UDAF)和用户自定义表生成函数(UDTF)。

5.1 函数简介

在数学中,一个函数是描述每个输入值对应唯一输出值的这种对应关系,符号通常为 f(x)。包含某个函数所有的输入值的集合被称作这个函数的定义域,包含所有的输出值的集合被称作值域。

计算机的函数是一个固定的程序段,或称其为一个子程序。它在实现固定运算功能的同时,还带有一个入口和一个出口。所谓的入口就是函数所带的各个参数,通过这个入口把函数的参数值代入子程序,供计算机处理。所谓出口,就是指函数的函数值,在计算机求得之后,由此口带回给调用它的程序。

Hive 函数,顾名思义,就是在写 Hive 语句时用到的一些函数,用于计算出用户想要得到的结果。Hive 函数有标准函数、聚合函数以及表生成函数等。

在 ETL 处理中,一个处理过程可能包含多个处理步骤。Hive 语言具有多种方式来将上一步骤的输入通过管道传递给下一个步骤,然后在一个查询中产生许多的输出。用户同样可以针对一些特定的处理过程编写自定义函数,允许用户扩展 HiveQL 的强大功能。

5.1.1 发现和描述函数

在 Hive 中,通常使用 UDF 来表示任意的函数,包括用户自定义的函数。在学习编写自定义函数 UDF 之前,先了解 Hive 中自带的 UDF。

在 Hive 中有查看数据库的命令,也有查看数据库中有哪些表的命令,那有没有命令可以查看 Hive 中有哪些函数呢?有的,SHOW FUNCTIONS 命令可以列出当前 Hive 会话中所加载的所有函数名称。

```
hive>SHOW FUNCTIONS;
```

```
OK
!
!=
%
&
*
```

通过 SHOW FUNCTIONS 命令查询出来的函数不只是上面的这些算术运算符,还有一些内置函数和用户的自定义函数。函数通常都有自己的使用文档,可以使用 DESCRIBE FUNCTION 命令展示对应函数的介绍。

【例 5-1】 查看 round 函数的简短使用文档。

```
hive>DESCRIBE FUNCTION round;
OK
round(x[, d]) - round x to d decimal places
Time taken: 0.111 seconds, Fetched: 1 row(s)
```

【例 5-2】 通过增加 EXTENDED 关键字可以查看更加详细的使用文档。

```
hive>DESCRIBE FUNCTION EXTENDED round;
OK
round(x[, d]) - round x to d decimal places
Example:
  > SELECT round(12.3456, 1) FROM src LIMIT 1;
  12.3'
Time taken: 0.057 seconds, Fetched: 4 row(s)
```

5.1.2 调用函数

如果要使用函数,只需要在查询中通过调用函数名,并传入所需要的参数就可以。某些函数需要指定特定的参数个数和参数类型,而其他函数可以传入一组参数,参数类型可以是多样的。和关键字一样,函数名也是保留的字符串。例如:

```
Hive>SELECT round(12.3456,1)FROM src LIMIT 1;
```

5.1.3 标准函数

用户自定义函数这个术语在狭义上的概念还表示以一行数据中的一列或多列数据作为参数,然后返回结果是一个值的函数。不仅如此,这些 UDF 还可以返回一个复杂的对象,例如 1.4.2 小节介绍过的 ARRAY、MAP 和 STRUCT。

在 Hive 中用到的函数包含了很多数学函数,例 5-1 使用的 round 函数就属于数学函数。表 5-1 中描述了 Hive 内置数学函数,用于处理单个列的数据。

表 5-1 Hive 内置数学函数

函 数	描 述	返回值类型
round(DOUBLE d)	输入 DOUBLE 类型的 d,返回 BIGINT 类型近似值	BIGINT
round(DOUBLE d,INT n)	输入 DOUBLE 类型的 d,返回保留 n 位小数的 DOUBLE 型近似值	DOUBLE
floor(DOUBLE d)	输入 DOUBLE 类型的 d,返回<=d 的最大 BIGINT 类型的值	BIGINT

续表

函　数	描　述	返回值类型
ceil(DOUBLE d) ceiling(DOUBLE d)	输入 DOUBLE 类型的 d，返回 $>=d$ 的最小 BIGINT 类型值	BIGINT
rand( ) rand(INT seed)	每行返回一个 DOUBLE 类型随机数，整数 $seed$ 是随机因子	DOUBLE
exp(DOUBLE d)	输入 DOUBLE 类型的 d，返回 DOUBLE 类型的 e 的 d 次幂	DOUBLE
ln(DOUBLE d)	返回以自然数为底 d 的对数，DOUBLE 类型	DOUBLE
log10(DOUBLE d)	返回以 10 为底 d 的对数，DOUBLE 类型	DOUBLE
log2(DOUBLE d)	返回以 2 为底 d 的对数，DOUBLE 类型	DOUBLE
log(DOUBLE base, DOUBLE d)	返回以 $base$ 为底 d 的对数，DOUBLE 类型，其中 $base$ 和 d 都是 DOUBLE 类型	DOUBLE
pow(DOUBLE d, DOUBLE p) power(DOUBLE d, DOUBLE p)	返回 d 的 p 次幂，DOUBLE 类型，其中 d 和 p 都是 DOUBLE 类型的	DOUBLE
sqrt(DOUBLE d)	返回 d 的平方根，其中 d 是 DOUBLE 类型	DOUBLE
bin(DOUBLE i)	返回二进制值 i 的 STRING 类型，其中 i 是 BIGINT 类型	STRING
hex(BIGINT i)	返回十六进制值 i 的 STRING 类型值，其中 i 是 BIGINT 类型	STRING
hex(STRING str)	返回十六进制表达的值 str 的 STRING 类型值	STRING
hex(BINARY b)	返回二进制表达的值 b 的 STRING 类型值（Hive 0.12.0 版本新增）	STRING
unhex(STRING i)	hex(STRING str)的逆方法	STRING
conv(BIGINT num,INT from_base,INT to_base)	将 BIGINT 类型的 num 从 $from\_base$ 进制转换成 to_base 进制，并返回 STRING 类型结果	STRING
conv(STRING num,INT from_base,INT to_base)	将 STRING 类型的 num 从 $from\_base$ 进制转换成 to_base 进制，并返回 STRING 类型结果	STRING
abs(DOUBLE d)	计算 DOUBLE 类型值 d 的绝对值，返回结果也是 DOUBLE 类型的	DOUBLE
pmod(INT i1,INT)	INT 类型值 $i1$ 对 INT 类型值 $i2$ 取模，结果也是 INT 类型	INT
pmod(DOUBLE d1,DOUBLE d2)	DOUBLE 类型值 $d1$ 对 DOUBLE 类型值 $d2$ 取模，结果也是 DOUBLE 类型的	DOUBLE
sin(DOUBLE d)	在弧度度量中，返回 DOUBLE 类型值 d 的正弦值，结果是 DOUBLE 类型的	DOUBLE
asin(DOUBLE d)	在弧度度量中，返回 DOUBLE 类型值 d 的反正弦值，结果是 DOUBLE 类型的	DOUBLE
cos(DOUBLE d)	在弧度度量中，返回 DOUBLE 类型值 d 的余弦值，结果是 DOUBLE 类型的	DOUBLE
acos(DOUBLE d)	在弧度度量中，返回 DOUBLE 类型值 d 的反余弦值，结果是 DOUBLE 类型的	DOUBLE
tan(DOUBLE d)	在弧度度量中，返回 DOUBLE 类型值 d 的正切值，结果是 DOUBLE 类型的	DOUBLE

续表

函　　数	描　　述	返回值类型
atan(DOUBLE d)	在弧度度量中,返回 DOUBLE 类型值 d 的反正切值,结果是 DOUBLE 类型的	DOUBLE
degrees(DOUBLE d)	将 DOUBLE 类型弧度值 d 转换成角度值,结果是 DOUBLE 类型的	DOUBLE
radians(DOUBLE d)	将 DOUBLE 类型角度值 d 转换成弧度值,结果是 DOUBLE 类型的	DOUBLE
positive(INT i)	返回 INT 类型值 i(其等价的有效表达式是\+i)	INT
positive(DOUBLE d)	返回 DOUBLE 类型值 d(其等价的有效表达式是\+d)	DOUBLE
negative(INT i)	返回 INT 类型值 i 的负数(其等价的有效表达式是−i)	INT
negative(DOUBLE d)	返回 DOUBLE 类型值 d 的负数(其等价的有效表达式是−d)	DOUBLE
sign(DOUBLE d)	如果 DOUBLE 类型值 d 是正数,则返回 FLOAT 类型值 1.0;如果 d 是负数,则返回−1.0;否则返回 0.0	FLOAT
e()	数学常数 e,也就是超越数,是 DOUBLE 类型值	DOUBLE
pi()	数学常数 pi,也就是圆周率,是 DOUBLE 类型值	DOUBLE

学习函数,就不得不说说类型转换问题。在进行数据类型转换时,floor、round 和 ceil ("向上取整")函数是首选的处理方式,而不是 5.1.4 小节要学习的 cast 类型转换函数。因为它们输入的是 DOUBLE 类型的值,而返回值是 BIGINT 类型的,也就是将浮点型数转换成整型了。

5.1.4　聚合函数

聚合函数是一类比较特殊的函数,其可以对多行进行一些计算,然后得到一个结果值。更确切地说,这是用户自定义聚合函数(UDAF)。所有的聚合函数、用户自定义函数和内置函数都统称为用户自定义聚合函数。表 5-2 列举了 Hive 的内置聚合函数。

表 5-2　Hive 的内置聚合函数

函　　数	描　　述	返回值类型
count(*)	计算总行数,包括含有 NULL 值的行	BIGINT
count(expr)	计算提供的 expr 表达式的值非 NULL 的行数	BIGINT
count(DISTINCT expr[, expr_.])	计算提供的 expr 表达式的值排重后非 NULL 的行数	BIGINT
sum(col)	计算指定行的值的和	DOUBLE
sum(DISTINCT col)	计算排重后值的和	DOUBLE
avg(col)	计算指定行的值的平均值	DOUBLE
avg(DISTINCT col)	计算排重后的值的平均值	DOUBLE
min(col)	计算指定行的最小值	DOUBLE
max(col)	计算指定行的最大值	DOUBLE

续表

函　数	描　述	返回值类型
variance(col),var_pop(col)	返回集合 col 中一组数值的方差	DOUBLE
var_samp(col)	返回集合 col 中一组数值的样本方差	DOUBLE
stddev_pop(col)	返回一组数值的标准偏差	DOUBLE
stddev_samp(col)	返回一组数值的标准样本偏差	DOUBLE
covar_pop(col1,col2)	返回一组数值的协方差	DOUBLE
covar_samp(col1,col2)	返回一组数值的样本协方差	DOUBLE
corr(col1,col2)	返回两组数值的相关系数	DOUBLE
percentile(BIGINT int_expr, p)	int_expr 在 p（范围：[0,1]）处对应的百分比，其中 p 是一个 DOUBLE 型数值	DOUBLE
percentile(BIGINT int_expr, ARRAY(p1[,p2]...))	int_expr 在 p（范围：[0,1]）处对应的百分比，其中 p 是一个 DOUBLE 型数组	ARRAY<DOUBLE>
percentile_approx(DOUBLE col,p[,NB])	col 在 p（范围：[0,1]）处对应的百分比，其中 p 是一个 DOUBLE 型数值，NB 是用于估计的直方图中的仓库数量（默认是 10000）	DOUBLE
percentile_approx(DOUBLE col,ARRAY(p1[,p2]...)[,NB])	col 在 p（范围：[0,1]）处对应的百分比，其中 p 是一个 DOUBLE 型数组，NB 是用于估计的直方图中的仓库数量（默认是 10000）	ARRAY<DOUBLE>
histogram_numeric(col,NB)	返回 NB 数量的直方图仓库数组。返回结果中的值 x 是中心，值 y 是仓库的高	ARRAY<STRUCT{'x','y'}>
collect_set(col)	返回集合 col 元素排重后的数组	ARRAY

表 5-3 中描述了 Hive 中其他的内置函数，这些函数用于处理字符串、Map、数组、JSON 和时间戳等。

表 5-3　Hive 中其他的内置函数

函　数	描　述	返回值类型
ascii(STRING s)	返回字符串 s 中首个 ASCII 字符的整数值	STRING
base64(BINARY bin)	将二进制 bin 转换成基于 64 位的字符串（Hive 0.12.0 版本新增）	STRING
binary(STRING s) binary(STRING b)	将输入的值转换成二进制值（Hive 0.12.0 版本新增）	BINARY
cast(<expr>as <type>)	将 expr 转换成 type 类型的。例如 cast('1'as BIGING)将会将字符串'1'转换成 BIGINT 数值类型。如果转换过程失败，则返回 NULL	返回类型就是 type 定义的类型
concat(BINARY s1,BINARY s2,...)	将二进制字节码按次序拼接成一个字符串（Hive 0.12.0 版本新增）	STRING
concat(STRING s1,STRING s2,...)	将字符串 s1,s2 等拼接成一个字符串。例如,concat('ab','cd')的结果是'abcd'	STRING
concat_ws(STRING separator, STRING s1,STRING s2,...)	和 concat 类似，不过是使用指定的分隔符进行拼接的	STRING

续表

函 数	描 述	返回值类型
concat_ws(STRING separator, BINARY s1, STRING s2,…)	和 concat 类似，不过是使用指定的分隔符进行拼接的(Hive 0.12.0 版本新增)	STRING
context_ngrams(array<array<string>>, array<string>, int k, int pf)	和 ngrams 类似，但是从每个外层数组的第二个单词数组来查找前 K 个字尾	ARRAY<STRUCT<STRING,DOUBLE>>
decode(BINARY bin, STRING charset)	使用指定的字符集 charset 将二进制值 bin 解码成字符串(支持的字符集有：'US_ASCII'、'ISO_8859_1'、'UTF-8'、'UTF-16BE'、'UTF-16LE'、'UTF-16')。如果任一参数输入为 NULL，则结果为 NULL(Hive 0.12.0 版本新增)	STRING
encode(STRING src, STRING charset)	使用指定的字符集 charset 将字符 src 串解码成二进制值(支持的字符集有：'US_ASCII'、'ISO_8859_1'、'UTF-8'、'UTF-16BE'、'UTF-16LE'、'UTF-16')。如果任一参数输入为 NULL，则结果为 NULL(Hive 0.12.0 版本新增)	BINARY
find_in_set(STRINGs, STRING commaSeparatedString)	返回在以逗号分隔的字符串中 s 出现的位置，如果没有找到就返回 NULL	INT
format_number(NUMBER x, INT d)	将数值 x 转换成'#,###,###.##'格式字符串，并保留 d 位小数。如果 d 为 0，那么输出值就没有小数点后面的值	STRING
get_json_object(STRING json_string, STRING path)	从给定路径上的 JSON 字符串中抽取出 JSON 对象，并返回这个对象的 JSON 字符串形式。如果输入的 JSON 字符串是非法的，则返回 NULL	STRING
in	例如，test in(val1,val2,…)，其表示如果 test 值等于后面列表中任一值，则返回 true	BOLLEAN
in_file(STRING s, STRING filename)	如果文件名为 filename 的文件中有完整一行数据和字符串 s 完全匹配，则返回 true	BOLLEAN
instr(STRING str, STRING substr)	查找字符串 str 中字符串 substr 第一次出现的位置	INT
length(STRING s)	计算字符串 s 的长度	INT
locate(STRING substr, STRING str[,INT pos])	查找字符串 str 中的 pos 位置后字符串 substr 第一次出现的位置	INT
lower(STRING s)	将字符串中所有字母转换成小写字母。例如，upper('hIvE')的结果是'hive'	STRING
lcase(STRING s)	和 lower()一样	STRING

续表

函　数	描　述	返回值类型
lpad（STRING s，INT Len，STRING pad）	从左边开始对字符串 s 使用字符串 pad 进行填充，最终达到 len 长度为止。如果字符串 s 本身长度比 len 大，那么多余的部分会被去除	STRING
ltrim(STRING s)	将字符串 s 前面出现的空格全部去除。例如 trim(' hive ')的结果是'hive '	STRING
ngrams（ARRAY＜ARRAY＜string＞＞，INT N，INT K，INT pf）	估算文件中前 K 个字尾。pf 是精度系数	ARRAY＜STRUCT＜STRING,DOUBLE＞＞
parse_url(STRING url，STRING partname[，STRING key])	从 URL 中抽取指定部分的内容。参数 url 表示一个 URL 字符串，参数 partname 表示要抽取的部分名称，其是大小写敏感的，可选的值有：HOST、PATH、QUERY、REF、PROTOCOL、AUTHORITY、FILE、USERINFO、QUERY：＜key＞。如果 partname 是 QUERY，那么还需要指定第三个参数 key。可以和表 5-4 中的 parse_url_tuple 对比下	STRING
printf（STRING format，Obj ... args）	按照 printf 风格格式化输出/输入的字符串（Hive 版本新增）	STRING
regexp_extract(STRING subject，STRING regex_pattern，STRING index)	抽取字符串 subject 中符合正则表达式 regex_pattern 的第 index 个部分的子字符串	STRING
regexp_replace（STRING s，STRING regex，STRING replacement）	按照 Java 正则表达式 regex 将字符串 s 中符合条件的部分替换成 replacement 所指定的字符串 a。如果 replacement 部分为空，那么符合正则的部门就会被去除。例如 regexp_replace('hive','[ie]','z')的结果是'hzvz'	STRING
repeat(STRING s,INT n)	重复输出 n 次字符串 s	STRING
reverse(STRING s)	反转字符串	STRING
rpad（STRING s，INT len，STRING pad）	从右边开始对字符串 s 使用字符串 pad 进行填充，最终达到 len 长度为止。如果字符串 s 本身长度比 len 大，那么多余的部分会被去除	STRING
rtim(STRING s)	将字符串 s 后面出现的空格全部去除。例如，trim(' hive ')的结果是' hive'	STRING
sentences(STRING s,STRING lang，STRING locale)	将输入字符串 s 转换成句子数组，每个句子又由一个单词数组构成。参数 lang 和 locale 是可选的，如果没有使用的，则使用默认的本地化信息	ARRAY＜ARRAY＜STRING＞＞
size(MAP＜K．V＞)	返回 MAP 中元素的个数	INT

续表

函 数	描 述	返回值类型
size(ARRAY<T>)	返回数组 ARRAY 的元素个数	INT
space(INT n)	返回 n 个空格	STRING
split（STRING s，STRING pattern）	按照正则表达式 pattern 分隔字符串 s，并将分隔后的部分以字符串数组的方式返回	ARRAY<STRING>
str_to_map（STRING s，STRING delim1，STRING delim2）	将字符串 s 按照指定分隔符转换成 Map，第一个参数是输入的字符串，第二个参数是键值对之间的分隔符，第三个分隔符是键和值之间的分隔符	MAP<STRING，STRING>
substr(STRING s，INT start_index) substring(STRING s，INT start_index)	对于字符串 s，截取从 start_index 位置开始到结尾的字符串，作为子字符串。例如，substr('abcdefgh',3)的结果是'cd efgh'	STRING
substr(BINARY s，INT start_index，INT length) substring(BINARY s,INT start_index，INT length)	对于二进制字节值 s，从 start 位置开始截取 length 长度的字符串，作为子字符串（Hive 0.12.0 新增）。例如，substr('abcdefgh',3,2)的结果是'cd'	STRING
translate（STRING input，STRING from，STRING to）	将 input 中出现在 from 中的字符替换为 to 中的字符串，如果任何参数为 NULL，结果为 NULL	STRING
trim(STRING s)	将字符串 s 前后出现的空格全部去除。例如 trim(' hive ')的结果是'hive'	STRING
unbase64(STRING str)	将基于 64 位的字符串 str 转换成二进制值（Hive 0.12.0 新增）	BINARY
upper（STRING s）ucase(STRING s)	将字符串中所有字母转换成大写字母。例如，upper('hIvE')的结果是'HIVE'	STRING
from_unixtime（BIGINT unixtime[，STRING format]）	将时间戳秒数转换成 UTC 时间，并用字符串表示，可以通过 format 规定的时间格式，指定输出的时间格式	STRING
unix_timestamp()	获取当前本地时区下的当前时间戳	BIGINT
unix_timestamp（STRING date）	输入的时间字符串格式必须是 yyyy-MM-dd HH：MM：SS，如果不符合则返回 0；如果符合，则将此时间字符串转换成 UNIX 时间戳。例如，unix_timestamp('2016-04-08 16：49：03')=1460105343503	BIGINT
unix_timestamp（STRING date，STRING pattern）	将指定时间字符串格式字符串转换成 UNIX 时间戳，如果格式不对，则返回 0。例如，unix_timestamp('2016-03-08','yyyy-MM-dd')=1460106565861	BIGINT
to_date(STRING timestamp)	返回时间字符串的日期部分。例如，to_date("1970-01-01 00：00：00")="1970-01-01"	STRING
year(STRING date)	返回时间字符串中的年份，并使用 INT 类型表示。例如，year("1970-01-01 00：00：00")=1970，year("1970-01-01")=1970	INT

续表

函数	描述	返回值类型
month(STRING date)	返回时间字符串中的月份，并使用 INT 类型表示。例如，month("1970-11-01 00：00：00")=11,month("1970-11-01")=11	INT
day(STRING date)dayofmonth(STRING date)	返回时间字符串中的天，并使用 INT 类型表示。例如，day("1970-01-01 00:00:00")=1, day("1970-01-01")=1	INT
hour(STRING date)	返回时间戳字符串中的小时，并使用 INT 类型表示。例如，hour("2016-04-01 00：00：00")=1,hour("1970-01-01")=1	INT
minute(STRING date)	返回时间字符串中的分钟数	INT
second(STRING date)	返回字符串中的秒数	INT
weekofyear(STRING date)	返回时间字符串位于一年中第几周内。例如，weekofyear("1970-11-01 00:00:00")=44, weekofyear("1970-11-01")=44	INT
datediff(STRING enddate, STRING startdate)	计算从开始时间 startdate 到结束时间 enddate 差了多少天。例如，datediff('2016-04-01','2016-04-08')=8	INT
date_add(STRING startdate, INT days)	为开始时间 startdate 增加了多少天。例如，date_add('2016-04-07',1)='2016-04-08'	STRING
date_sub(STRING startdate, INT days)	为开始时间 startdate 减少 days 天。例如，date_add('2016-04-08',1)='2016-04-07'	STRING
from_utc_timestamp(TIMESTAMP timestamp, STRING timezone)	如果给定的时间戳并非 UTC,则将其转化成指定的时区下的时间戳（Hive 0.8.0 版本新增）	TIMESTAMP
to_utc_timestamp(TIMESTAMP timestamp, STRING timezone)	如果给定的时间戳是指定的时区下的时间戳，则将其转化成 UTC 下的时间戳（Hive 0.8.0 版本新增）	TIMESTAMP

Hive 函数中的类型转换函数是 cast()，用户可以使用 cast() 函数对指定的值进行显示的类型转换。回想 3.2 节创建过的 sougou_20111230 表中，rank 列是使用 INT 数据类型的。假设用户现在需要将此表中的前 3 条的记录的 rank 字段的数据类型转换成 DOUBLE 数据类型，应该怎样使用 cast() 函数呢？

```
hive>SELECT uid,keyword,rank,cast(rank as DOUBLE)
    >FROM sougou_20111230 LIMIT 3;
OK
698956eb07815439fe5f46e9a4503997        youku    1        1.0
f31f594bd1f3147298bd952ba35de84d        12306.cn    1        1.0
596444b8c02b7b30c11273d5bbb88741        pissing videos 1        1.0
Time taken: 1.901 seconds, Fetched: 3 row(s)
```

注意：如果被转换数据类型的字段的值不合法，Hive 会返回 NULL。将浮点数转换成整数的推荐方式是使用 round() 或者 floor() 函数，而不推荐使用 cast() 函数进行数据转换。

get_json_object 函数的使用实例如下：

```
hive> SELECT get_json_object('{"name":"xiaoming","age":"18"}','$.age')
    >FROM sougou_20111230 LIMIT 3;
OK
18
18
18
Time taken: 1.032 seconds, Fetched: 3 row(s)
```

5.1.5 表生成函数

与聚合函数"相反的"一类函数是表生成函数。和其他函数类别一样,用户自定义的和内置的表生成函数统称为用户自定义表生成函数。表生成函数可以将单列扩展成多列或者多行,它弥补了用户自定义函数(UDF)无法返回多行或多列的缺陷。表 5-4 列举了 Hive 内置的表生成函数。

表 5-4 Hive 内置的表生成函数

函 数	描 述	返回值类型
explode(ARRAY array)	返回 0 到多行结果,每行都对应输入的 array 数组中的一个元素	N 行结果
explode(MAP map)	返回 0 到多行结果,每行对应每个 map 键-值对,其中一个字段是 map 的键,另一个字段对应 map 的值(Hive 0.8.0 版本新增)	N 行结果
explode(ARRAY\<TYPE\>a)	对于 a 中的每个元素,explode()会生成一行记录包含这个元素	数组的类型
inline(ARRAY\<STRUCT [,STRUCT]\>)	将结构体数组提取出来并插入表中(Hive 0.10.0 版本新增)	结果插入表中
json_tuple(STRING jsonStr,p1, p2,…,pn)	本函数可以接受多个标签名称,对输入的 JSON 字符串进行处理,这个 get_json_object 与 UDF 类似,不过更高效,其通过一次调用就可以获得多个键值	TUPLE
parse_url_tuple(url,Partname1, partname2,…,partnamen)	其中 $n>=1$,从 URL 中解析出 n 个部分信息。其输入参数是:URL,以及多个要抽取的部分的名称。所有输入的参数的类型都是 STRING。部分名称是大小写敏感的,而且不应该包含空格:HOST、PATH、QUERY、REF、PROTOCOL、AUTHORITY、FILE、USERINFO、QUERY:\<KEY_NAME\>	TUPLE
stack(INT n,col1,…,colm)	把 m 列转换成 n 行,每行有 m/n 个字段。其中,n 必须是个常数	N 行结果

实例:假设有一张表 test,表中存放了一个数组[1,2,3,4],用 array 函数将一列输入转换成一个数组输出。

```
SELECT array(1,2,3,4)FROM test;
```

输出为:

[1,2,3,4]

用 explode()函数以 array 类型数据作为输入,然后对数组中的数据进行迭代,返回多行结果,每行一个数组的元素值。

```
SELECT explode(array(1,2,3,4))FROM test;
```

输出为:

1
2
3
4

5.2 用户自定义函数 UDF

本节主要是通过一个案例来学习如何写一个 UDF 实例——通过日期计算其星座的 UDF。首先要创建一个用户信息表,表中的一个字段存储的是每个用户的生日。通过这些信息计算出每个人所属的星座。

首先,在 Hive 中创建一个表,用于存储用户的相关信息。

```
hive>CREATE TABLE IF NOT EXISTS userdata(
    >name STRING,
    >email STRING,
    >bday STRING,
    >ip STRING,
    >sex STRING,
    >anum INT)
    >ROW FORMAT DELIMITED
    >FIELDS TERMINATED BY ','
    >LOCATION '/data/userdata';
```

然后,准备数据。在 Hadoop 本地文件系统的 Home 目录下创建一个文件 userdata,用于存储用户信息,表 userdata 的内容如下:

```
wangwenyi,wang@media6degrees.com,25-12-1983,209.191.200,M,10
zhaojiu@test.net,10-10-2002,10.10.10.1,M,50
lius,8983083@qq.com,4-5-1994,64.64.5.1,F,2
```

把 userdata 中的数据加载到 Hive 表 userdata 中:

```
hive> LOAD DATA LOCAL INPATH '/home/zkpk/userdata' INTO TABLE userdata;
Loading data to table sogou.userdata
Table sogou.userdata stats: [numFiles=0, totalSize=0]
OK
Time taken: 0.399 seconds
```

下面是这个 UDF 的 Java 代码。代码中@Description(…)表示的是 Java 的所有注解,是可以选择的。注解中注明了函数的文档说明,用户需要通过注解来阐明自定义的 UDF 的使用方法和例子。这样当用户通过 DESCRIBE FUNCTION…命令查看该函数时,注解中的_FUNC_字符串将会被替换为这个函数定义的"临时"函数名称。

```
package com0411;
```

```java
import java.text.ParseException;
import java.text.SimpleDateFormat;
import java.util.Date;
import org.apache.hadoop.hive.ql.exec.UDF;

/* @Description(name="star",
value="_FUNC_(date) - from the input date String "+"or separate month and day
arguments,returns the sign of the Star.",
extended="Example:\n"
+" >SELECT _FUNC_(date_string)FROM src;\n"
+" >SELECT _FUNC_(month,day)FROM src;")
*/

public class StarSignUDF extends UDF{
    private SimpleDateFormat df;
    public StarSignUDF(){
        df=new SimpleDateFormat("MM-dd-yyyy");
    }
    public String evaluate(Date bday){
        return this.evaluate(bday.getMonth(),bday.getDay());
    }
    public String evaluate(String bday){
        Date date=null;
        try {
            date=df.parse(bday);
        } catch(Exception e){
            return null;
        }
        return this.evaluate(date.getMonth()+1,date.getDate());
    }
    public String evaluate(Integer month,Integer day){
        if(month==1){
            if(day<20){
                return "Capricorn";
            }else {
                return "Aquarius";
            }
        }
        if(month==2){
            if(day<19){
                return "Aquarius";
            }else {
                return "Pisces";
            }
        }
        if(month==3){
            if(day<21){
                return "Pisces";
            }else {
                return "Aries";
            }
```

```
            }
            if(month==4){
                if(day<20){
                    return "Aries";
                }else{
                    return "Taurus";
                }
            }
            if(month==5){
                if(day<21){
                    return "Taurus";
                }else{
                    return "Gemini";
                }
            }
            if(month==6){
                if(day<22){
                    return "Gemini";
                }else{
                    return "Cancer";
                }
            }
            if(month==7){
                if(day<23){
                    return "Cancer";
                }else{
                    return "Leo";
                }
            }
            if(month==8){
                if(day<23){
                    return "Leo";
                }else{
                    return "Virgo";
                }
            }
            if(month==9){
                if(day<23){
                    return "Virgo";
                }else{
                    return "Libra";
                }
            }
            if(month==10){
                if(day<24){
                    return "Libra";
                }else{
                    return "Scorpio";
                }
            }
```

```
            if(month==11){
                if(day<23){
                    return "Scorpio";
                }else {
                    return "Sagittarius";
                }
            }
            if(month==12){
                if(day<22){
                    return "Sagittarius";
                }else {
                    return "Capricorn";
                }
            }
            return null;
    }
}
```

从上面编写的 UDF 可以看出，一个 UDF 需要继承 UDF 类并实现 evaluate()方法。在查询执行过程中，查询中每个应用到这个函数的地方都会被这个类进行实例化。对于每行输入都会调用到 evaluate()方法，evaluate()处理后的值会返回给 Hive。当然，用户是可以重载 evaluate 方法的，Hive 会像 Java 的方法重载一样，自动选择匹配的方法。

1. UDF 函数编译、打包

如果想在 Hive 中使用自己编写的 UDF，那么需要将 Java 代码进行编译，编译无误后把它打包成一个 JAR 文件。然后，在 Hive 会话中将这个 JAR 文件加入类的路径下，通过 CREATE FUNCTION 语句定义这个 Java 类函数，方便以后使用它。具体操作步骤如下：

```
hive>ADD JAR /home/zkpk/star.jar;
hive>CREATE TEMPORARY FUNCTION star
    >AS 'com0411.StarSignUDF';
OK
```

注意：JAR 文件路径是不需要用引号括起来的，这个路径指的是存储 JAR 文件的全路径。在使用 CREATE FUNCTION 语句定义 star 函数时，AS 子句后面单引号里面的内容是 UDF 的包名.类名。

2. UDF 函数的调用

使用刚才创建的 UDF 函数，使用表 userdata 中的 bday 字段，计算出用户的星座：

```
hive> SELECT name,bday,starSign(bday) FROM userdata;
OK
wangwenyi        25-12-1983       Capricorn
zhaojiu 10-10-2002       Libra
lius    4-5-1994         Aries
Time taken: 0.29 seconds, Fetched: 3 row(s)
```

刚才使用的 starSign 函数可以像其他的函数一样的使用了。

在创建函数时用到的关键字 TEMPORARY，表示当前会话中声明的函数只会在当前会话中有效。所以，用户需要在每个会话中都增加 JAR，然后创建函数。如果用户频繁地使用同一个 JAR 文件和函数,那么需要将相关的语句增加到 $HOME/.hiverc 文件中。

3. UDF 函数的删除

当用户使用完自定义 UDF 后,可以通过以下语句删除此函数。

```
hive> DROP TEMPORARY FUNCTION starSign;
OK
Time taken: 0.056 seconds
```

5.3 用户自定义聚合函数 UDAF

在之前已经学习了自定义函数的相关概念及一些方法,UDF 只能实现一进一出的操作,如果要实现多进一出的聚合类操作,则要用到用户自定义聚合函数 UDAF。用户自定义聚合函数最大的特点就是其可以对多行进行一些计算,然后得到一个结果值。UDAF 的具体用法如下:

(1) 以下两个包是必须导入的。

```
import org.apache.hadoop.hive.ql.exec.UDAF
import org.apache.hadoop.hive.ql.exec.UDAFEvaluator
```

(2) 函数类需要继承 UDAF 类,内部类 Evaluator 实现 UDAFEvaluator 接口。

(3) Evaluator 需要实现 init、iterate、terminatePartial、merge、terminate 等函数。

① init 函数实现接口 UDAFEvaluator 的 init 函数。

② iterate 接收传入的参数,并进行内部的轮转。其返回类型为 boolean。

③ terminatePartial 无参数,其为 iterate 函数轮转结束后,返回轮转数据,terminatePartial 类似于 Hadoop 的 Combiner。

④ merge 接收 terminatePartial 的返回结果,进行数据 merge 操作,其返回类型为 boolean。

⑤ terminate 返回最终的聚集函数结果。

下面用一个非常典型的实例来学习怎样写一个 UDAF,该函数要实现这样一个功能:在表 customers 中,查询 age 字段,找出最小的年龄。

首先创建一个 customers 表,其中包括的字段是姓名,性别和年龄,用'\t'键分隔。

```
hive>CREATE TABLE customers(
    >name STRING,
    >sex STRING,
    >age INT)
    >ROW FORMAT DELIMITED
    >FIELDS TERMINATED BY '\t'
    >LOCATION '/data/customers';
```

把数据加载在 Hive 表中。

```
hive>LOAD DATA LOCAL INPATH '/home/zkpk/customers' INTO TABLE customers;
```

编写一个 UDAF,实现找出最小年龄功能,代码如下:

```java
package com0411;
import org.apache.hadoop.hive.ql.exec.UDAF;
import org.apache.hadoop.hive.ql.exec.UDAFEvaluator;
import org.apache.hadoop.io.IntWritable;

public class MinNumUDAF extends UDAF{

    public static class MinNumIntUDAFEvaluator implements UDAFEvaluator {
        private IntWritable result;

        public void init(){
            result=null;
        }

        public boolean iterate(IntWritable value){
            if(value ==null){
                return true;
            }
            if(result ==null){
                result=new IntWritable(value.get());
            } else {
                result.set(Math.min(result.get(), value.get()));
            }
            return true;
        }

        public IntWritable terminatePartial(){
            return result;
        }

        public boolean merge(IntWritable other){
            return iterate(other);
        }

        public IntWritable terminate(){
            return result;
        }
    }
}
```

1. UDAF 函数的编译、打包

如果想在 Hive 中使用自己编写的 UDAF,那么需要将 Java 代码进行编译,编译无误后把它打包成一个 JAR 文件。然后,在 Hive 会话中将这个 JAR 文件加入类的路径下,通过 CREATE FUNCTION 语句定义这个 Java 类函数,方便以后使用它。

把编写的 UDAF 打成一个 JAR 包,然后添加该 jar 包到类路径下:

hive>ADD JAR /home/zkpk/customer.jar;

2. UDAF 函数的声明

把 UDAF 打成一个 JAR 包后,还需要在 Hive 中为它声明一个函数,这样在使用此函

数时,此函数才会有效。声明函数的命令如下:

```
hive>CREATE TEMPORARY FUNCTION minNum
    >AS 'com0411.MinNumUDAF';
```

3. UDAF 函数的调用

用刚刚声明的 minNum 函数,计算出表 customers 中的最小年龄。

```
hive>SELECT minNum(age)FROM customers;
2016-04-11 02:40:30,174 Stage-1 map = 0%,  reduce = 0%
2016-04-11 02:41:12,684 Stage-1 map = 100%,  reduce = 0%, Cumulative CI
2016-04-11 02:41:28,667 Stage-1 map = 100%,  reduce = 100%, Cumulative
MapReduce Total cumulative CPU time: 18 seconds 570 msec
Ended Job = job_1460336870082_0004
MapReduce Jobs Launched:
Stage-Stage-1: Map: 1  Reduce: 1   Cumulative CPU: 18.57 sec   HDFS Re
: 3 SUCCESS
Total MapReduce CPU Time Spent: 18 seconds 570 msec
OK
18
```

5.4 用户自定义表生成函数 UDTF

UDTF(User-Defined Table-Generating Functions)用于解决输入一行输出多行(One-to-many maping)的需求。用户编写自己需要的 UDTF,需要继承 org.apache.hadoop.hive.ql.udf.generic.GenericUDTF 类,实现 initialize、process、close 三个方法。

UDTF 首先会调用 initialize 方法,此方法返回 UDTF 的返回行的信息(返回个数、类型)。初始化完成后,会调用 process 方法,真正的处理过程在 process 函数中。在 process 中,每一次 forward()调用产生一行;如果产生多列,可以将多个列的值放在一个数组中,然后将该数组传入 forward()函数。最后调用 close()方法,对需要清理的方法进行清理。

下面实现一个 UDTF,用于切分"key:value;key:value;"这种字符串,返回结果为 key、value 两个字段。

```java
import java.util.ArrayList;

import org.apache.hadoop.hive.ql.udf.generic.GenericUDTF;
import org.apache.hadoop.hive.ql.exec.UDFArgumentException;
import org.apache.hadoop.hive.ql.exec.UDFArgumentLengthException;
import org.apache.hadoop.hive.ql.metadata.HiveException;
import org.apache.hadoop.hive.serde2.objectinspector.ObjectInspector;
import org.apache.hadoop.hive.serde2.objectinspector.ObjectInspectorFactory;
import org.apache.hadoop.hive.serde2.objectinspector.StructObjectInspector;
import org.apache.hadoop.hive.serde2.objectinspector.primitive.PrimitiveObjectInspectorFactory;

public class ExplodeMap extends GenericUDTF {

    @Override
    public void close()throws HiveException {
        //TODO Auto-generated method stub
    }
```

```java
@Override
public StructObjectInspector initialize(ObjectInspector[] args)
        throws UDFArgumentException {
    if(args.length !=1){
        throw new UDFArgumentLengthException(
                "ExplodeMap takes only one argument");
    }
    if(args[0].getCategory()!=ObjectInspector.Category.PRIMITIVE){
        throw new UDFArgumentException(
                "ExplodeMap takes string as a parameter");
    }

    ArrayList<String> fieldNames=new ArrayList<String>();
    ArrayList<ObjectInspector> fieldOIs=new ArrayList<ObjectInspector>();
    fieldNames.add("col1");

    fieldOIs.add(PrimitiveObjectInspectorFactory.javaStringObjectInspector);
    fieldNames.add("col2");

    fieldOIs.add(PrimitiveObjectInspectorFactory.javaStringObjectInspector);

    return ObjectInspectorFactory.getStandardStructObjectInspector(
            fieldNames, fieldOIs);
}

@Override
public void process(Object[] args)throws HiveException {
    String input=args[0].toString();
    String[] test=input.split(";");
    for(int i=0; i<test.length; i++){
        try {
            String[] result=test[i].split(":");
            forward(result);
        } catch(Exception e){
            continue;
        }
    }
}
```

使用方法同样可以按照前面的 UDF 和 UDAF 的方法,打成 jar 包后加入 hive 类路径中,然后声明该函数即可使用。

5.5 UDF 的标注

在 5.2 节用户自定义函数 UDF 中提到过 Description 标注,它用于在运行时为 Hive 方法提供说明文档。当然,这部分也可以省略,加上注解主要是为了以后更加方便地使用它。

除此之外,UDF 中还存在有其他的标注,正确使用这些标注可以使函数更加简单,甚至有时候对于某些 Hive 查询,可以提高执行效率。接下来,分别介绍定数性标注、状态性标注和唯一性标注。

```
public @interface UDFType{
    boolean deterministic()default true;
    boolean stateful()default false;
    boolean distinctLike()default false;
}
```

5.5.1 定数性标注(deterministic)

定数,顾名思义,是不会变的数。在默认情况下,大多数的查询都是满足定数性的,因为它们本身就具有定数性。

但是,rand()函数是个例外。如果一个 UDF 是非定数的,那么就不会包含在分区剪裁中。下面展示使用 rand()函数,查询一个非定数性的例子。

```
SELECT * FROM tablename WHERE rand()<0.01;
```

如果 rand()是定数的,那么结果只会在计算阶段计算一次。然而上面这个例子中,rand()的查询是非定数的,所以对于每行数据,rand()的值都需要重新计算一次。

5.5.2 状态性标注(stateful)

几乎所有的 UDF 默认都是有状态性的。状态性标注的适用的情况如下:

(1) 有状态的 UDF 只能用在 SELECT 语句后,而不能用到其他语句后,如 WHERE、ON、ORDER、GROUP 等语句。

(2) 当一个查询语句中存在有状态的 UDF 时,那么隐含的信息就是 SELECT 将会和 TRANSFORM(例 DISTRIBUTE、CLUSTER、SORT 语句)进行类似的处理,然后会在对应的 reducer 内部执行,以保证结果能够满足预期。

(3) 如果状态标记 stateful 设置值为 true,那么这个 UDF 同样应该作为非定数性的(即使这时定数性标记 deterministic 的值是显式设置为 true 的)。

5.5.3 唯一性标注(distinctLike)

有些函数即使输入的列是非排重值,它的结果也是类似于使用了 DISTINCT 进行了排重操作,这类场景可以定义为唯一性,如函数中的 max 和 min 函数。

本章小结

本章主要介绍了 Hive 中的各类函数。

(1) 运用命令的形式,查看 Hive 中已经存在的函数,并且运用命令查看函数的使用文档。

(2) 学习了常用的标准函数、聚合函数和表生成函数,了解不常用的函数。

(3) 学习了通过日期计算其星座的 UDF。

(4) 学习了 UDAF 和 UDTF 的案例,前者的功能是多行输入一行输出,后者的功能是一行输入多行输出。

习　题

1. 选择题

(1) 在数据类型转换操作中,(　　)是描述函数 floor(DOUBLE d)的。
　　A. 将 expr 转换成 type 类型。
　　B. 输入 DOUBLE 类型 d,返回 BIGINT 类型的近似值
　　C. d 是 DOUBLE 类型的,返回>=d 的最大 BIGINT 型值
　　D. d 是 DOUBLE 类型的,返回<=d 的最大 BIGINT 型值

(2) 下列函数中,(　　)计算对列 col 的值排重后非 NULL 的行数。
　　A. count(col)　　　　　　　　　　　B. sum(col)
　　C. count(DISTINCT col)　　　　　　D. sum(DISTINCT col)

(3) get_json_object 函数的含义是从给定路径上的 JSON 字符串中抽取出 JSON 对象,并返回这个对象的 JSON(　　)形式。如果输入的 JSON 字符串是非法的,则返回 NULL。
　　A. 数组　　　　B. 字符串　　　　C. 整数　　　　D. 字符

(4) 下列函数中,不可以返回 N 行结果的是(　　)。
　　A. explode(ARRAY array)　　　　　B. explode(MAP map)
　　C. stack(INT n,col1,…,colm)　　　D. explode(ARRAY<TYPE> a)

(5) (　　)不属于 UDF 标注。
　　A. determined　　　　　　　　　　B. deterministic
　　C. stateful　　　　　　　　　　　　D. distinctLike

2. 问答题

(1) 在类型转换操作中,可以用到的函数有哪些?分别怎样使用它们?

(2) 谈谈对 get_json_object 函数的理解。

(3) 什么是 UDF?编写一个在字符串中去除空格的 UDF。

(4) 什么是 UDAF?简述 UDAF 的适用场景。

(5) 简述 UDTF 及运用一个 UDTF 的全过程(提示:从数据准备到结果展示)。

第 6 章

认识 Pig

本章摘要

Pig 为大型数据集的处理提供了更高层次的抽象。Pig 是一种编程语言，简化了 Hadoop 常见的工作任务，可实现加载数据、表达转换数据和存储最终结果。

本章将从 Pig 的基本概念、应用场景、设计思想、发展简史讲起，然后详细地讲解 Pig 的安装及运行，让读者在学习本章的内容后能对 Pig 有一个初步的认识。

6.1 初识 Pig

6.1.1 Pig 是什么

Pig 是 Apache 平台下的一个免费开源项目，它提供了一个基于 Hadoop 的并行执行数据流处理的引擎，为大型数据集的处理提供了更高层次的抽象。很多时候数据的处理需要多个 MapReduce 过程才能实现，使数据处理过程与该模式匹配可能很困难。有了 Pig 就能够使用更丰富的数据结构，这些数据结构往往都是多值和嵌套的，Pig 还提供了一套更强大的数据交换操作，包括在 MapReduce 中被忽略的链接操作。Pig 包括两部分。

（1）Pig Latin。它提供的 SQL-LIKE 语言叫 Pig Latin，该语言的编译器会把类 SQL 的数据分析请求转换为一系列经过优化处理的 MapReduce 运算。Pig Latin 是一个相对简单的语言，它可以执行语句。一条语句就是一个操作，它需要输入一些内容（比如代表一个元组集的包），并发出另一个包作为其输出。一个包就是一个关系，与表类似，可以在关系数据库中找到它。元组代表行，每个元组都由字段组成。

（2）用于运行 Pig Latin 程序的执行环境。当前包括单 JVM 中的本地执行环境和 Hadoop 集群上的分布式执行环境。

6.1.2 Pig 的应用场景

目前 Pig 主要用于离线数据的批量处理应用场景。随着 Pig 的发展，处理数据的速度会不断地提升，这可能依赖于 Pig 底层的执行引擎。例如，Pig 通过指定执行模式，可以使用 Hadoop 的 MapReduce 计算引擎来实现数据处理，也可以使用基于 Tez 的计算引擎来实现（Tez 是为了绕开 MapReduce 多阶段 Job 写磁盘而设计的 DAG 计算引擎，性能比 MapReduce 要快）。另外，根据 Pig 未来的发展路线图，以后可能会基于 Storm 或 Spark 计算平台实现底层计算引擎，那样速度会有极大的提升。

Pig 可以非常轻松地对 TB 级别海量的数据进行查询,并且这些海量的数据都是非结构化的数据。例如,一堆文件可能是 log4j 输出日志,又存放于跨越多个计算机的多个磁盘上,用于记录上千台在线服务器的健康状态日志、交易日志、IP 访问记录和应用服务日志等,通常需要统计或者抽取这些记录,或者查询异常记录。当需要对这些记录形成一些报表、将数据转化为有价值的信息时,查询会较为复杂,此时类似 MySQL 这样的产品就很难满足对速度、执行效率的需求,而用 Apache 的 Pig 就可以帮助用户去实现这样的目标。

需要注意的是,Pig 并不适合所有的数据处理任务。和 MapReduce 一样,它是为数据批处理而设计的。如果需要处理的是 GB 或者 TB 数量级的数据,那么 Pig 是个不错的选择;如果想执行的查询只涉及一个大型数据集中的一小部分数据,尤其对于那些需要写单条或者少量记录,或者查询随机序列下的多条不同记录的任务,Pig 的实现不会很好,因为它要扫描整个数据集或其中的很大一部分。

6.1.3 Pig 的设计思想

Pig 团队发布的项目设计声明表明了 Pig 的设计思想。

1. Pig 什么都吃

不管数据是否有元数据,Pig 都可以操作。不管数据是关系型的、嵌套型的,还是非结构化的,Pig 都同样可以操作。而且,它可以很容易地进行扩展,不仅可以操作文件,还可以操作 key/value 类型的存储,甚至数据库等数据源。

2. Pig 无处不在

Pig 期望成为一种并行数据处理语言。它不会局限于一种特殊的并行处理框架,首先是基于 Hadoop 之上实现的,但是期望并非只能在 Hadoop 平台上使用。

3. Pig 是家畜

Pig 被设计为可以让用户很容易地控制和修改的语言。

Pig 允许用户随时整合加入其他代码,因此目前它支持用户自定义字段类型转换函数、用户自定义聚合方法函数和用户定义条件式函数。这些函数可以使用 Java 来写,也可以使用最终可以编译成 Java 代码的脚本语言(Jython)编写。Pig 支持存储函数和用户定义的加载。Pig 通过自己的 stream 命令和需要 Mapreduce 相关的 JAR 包的 mapreduce 命令可执行外部的执行命令。Pig 也允许用户为自己的特定使用场景提供一个用户自定义的分区方法函数,使他们执行的任务在 reduce 阶段可以达到负荷均衡。

Pig 的优化器可以将 Pig Latin 脚本中的操作过程进行重新排列,从而达到更好的性能,如将 MapReduce 任务进行合并等。若优化过程是不必要的,用户可以很容易地将最优控制器关闭,这样执行过程就不会发生改变。

4. Pig 会飞

Pig 处理数据速度快。Pig 团队会持续地优化性能,同时不会增加一些使 Pig 显得较重而降低性能的新功能。

使用 Pig 来操作 Hadoop 处理海量数据是非常简单的。如果没有 Pig,就得手写 MapReduce 代码,这可是一件非常烦琐的事,因为 MapReduce 的任务职责非常明确——清洗数据需要一个 job,处理需要一个 job,过滤需要一个 job,统计需要一个 job,排序需要一个 job,编写 DAG(带先后顺序依赖的)作业很不方便,每次即使只改动很小的一个地方,也要重新编译整个job，然后打成jar提交到Hadoop集群上运行，调试还很困难。所以，在现在的大互联网公司或者是电商公司里，很少有纯写MapReduce来处理各种任务的，基本上都会使用一些工具或开源框架来操作。

Pig就是为了屏蔽MapReduce开发的烦琐细节，为用户提供Pig Latin这样近SQL语言处理能力，使用户可以更方便地处理海量数据。Pig将SQL语句翻译成MR的作业的集合，并通过数据流的方式将其组合起来。

6.1.4　Pig的发展简史

Pig最早是雅虎公司的一个基于Hadoop的并行处理架构。Yahoo!的科学家们设计了Pig，并给出了一个原型实现。最初Yahoo!的Hadoop使用者开始采用Pig，之后一个开发工程师团队开始接手Pig的最初原型，并将这个原型开发成一个达到产品级别的可用产品。那么，雅虎公司主要使用Pig来干什么呢？

（1）吸收和分析用户的行为日志数据（点击流分析、搜索内容分析等），改进匹配和排名算法，以提高检索和广告业务的质量。

（2）构建和更新search index。web-crawler抓取了的内容是一个流数据的形式，这包括去冗余、链接分析、内容分类、基于点击次数的受欢迎程度计算（PageRank），最后建立倒排表。

（3）处理半结构化数据订阅（data seeds）服务。包括deduplcaitin（去冗余）、geographic location resolution（地理位置分辨率）和named entity recognition（命名实体识别）。

2007年秋天，Pig通过Apache孵化器进行开源。2008年9月，Pig的第一个发布版本出现了，同年晚些时候，Pig从孵化器中毕业，正式提升为Apache Hadoop项目的一个子项目。

2009年早期，其他公司在他们的数据处理中开始使用Pig，亚马逊也将Pig加入它的弹性MapReduce服务中。2009年年末，Yahoo!公司所运行的Hadoop任务中有一半是Pig任务。在2010年，Pig发展持续增长，这一年Pig从Hadoop的子项目中脱离出来，自己成为一个最高级别的Apache项目。

6.2　安装、运行Pig

本书所用Pig的版本为0.13.0，官网下载地址：http://www.apache.org/dyn/closer.cgi/pig。

6.2.1　安装Pig

Pig的安装条件。

（1）Pig有Local模式和MapReduce模式两种运行模式。如果需要作业在分布式环境下运行，则需要安装Hadoop，否则可以选择不安装。

（2）Java环境对于Pig来说是必需的。

在安装Pig之前，要求已经成功安装了Hadoop，配置好了JDK，并已经正常启动Hadoop。

以将 Pig 安装在 HadoopMaster 节点上为例(以下操作都在 HadoopMaster 节点上进行)。
(1) 使用 zkpk 用户(密码：zkpk)，切换 zkpk 用户的命令是：

[zkpk@master ~]$ su - zkpk

输入密码：

zkpk

(2) 下载好后解压：

[zkpk@master ~]$ tar - zvxf pig-0.13.0-tar.gz

(3) 执行 ls -l 命令，会看到如下内容，这些内容是 Pig 包含的文件。

```
zkpk@master:~/pig-0.13.0                                    _ □ ×
File  Edit  View  Search  Terminal  Help
[zkpk@master pig-0.13.0]$ ls -l
total 34856
drwxr-xr-x. 2 zkpk zkpk       4096 Aug 30 22:26 bin
-rw-rw-r--. 1 zkpk zkpk      87837 Jun 29  2014 build.xml
-rw-rw-r--. 1 zkpk zkpk     163510 Jun 29  2014 CHANGES.txt
drwxr-xr-x. 2 zkpk zkpk       4096 Aug 30 22:26 conf
drwxr-xr-x. 3 zkpk zkpk       4096 Aug 30 22:26 contrib
drwxr-xr-x. 6 zkpk zkpk       4096 Aug 30 22:26 docs
drwxr-xr-x. 2 zkpk zkpk       4096 Aug 30 22:26 ivy
-rw-rw-r--. 1 zkpk zkpk      23441 Jun 29  2014 ivy.xml
drwxr-xr-x. 3 zkpk zkpk       4096 Aug 30 22:26 lib
drwxr-xr-x. 3 zkpk zkpk       4096 Jun 29  2014 lib-src
drwxr-xr-x. 2 zkpk zkpk       4096 Aug 30 22:26 license
-rw-rw-r--. 1 zkpk zkpk      11358 Jun 29  2014 LICENSE.txt
-rw-rw-r--. 1 zkpk zkpk       2125 Jun 29  2014 NOTICE.txt
-rw-rw-r--. 1 zkpk zkpk   17826437 Jun 29  2014 pig-0.13.0-h1.jar
-rw-rw-r--. 1 zkpk zkpk    8748886 Jun 29  2014 pig-0.13.0-withouthadoop-
h1.jar
-rw-rw-r--. 1 zkpk zkpk    8756785 Jun 29  2014 pig-0.13.0-withouthadoop-
h2.jar
-rw-rw-r--. 1 zkpk zkpk       1307 Jun 29  2014 README.txt
-rw-rw-r--. 1 zkpk zkpk       2564 Jun 29  2014 RELEASE_NOTES.txt
drwxr-xr-x. 2 zkpk zkpk       4096 Jun 29  2014 scripts
drwxr-xr-x. 4 zkpk zkpk       4096 Jun 29  2014 shims
drwxr-xr-x. 8 zkpk zkpk       4096 Aug 30 22:26 src
drwxr-xr-x. 9 zkpk zkpk       4096 Aug 30 22:26 test
drwxr-xr-x. 5 zkpk zkpk       4096 Aug 30 22:26 tutorial
[zkpk@master pig-0.13.0]$
```

6.2.2 运行 Pig

运行 Pig 之前需要先配置文件，在 zkpk 用户下编辑以下文件。

[zkpk@zkpk ~]$ vim .bash_profile

把下面命令加入配置文件的末尾。

export PIG_HOME=/home/zkpk/pig-0.13.0
export PATH=$PATH:$PIG_HOME/bin:$PIG_HOME/conf
export PIG_CLASSPATH=$HADOOP_HOME/etc/hadoop/

使用下面命令使文件生效。

[zkpk@zkpk ~]$ source .bash_profile

再切换到普通用户下,执行以下命令,查看是否安装成功。

[zkpk@zkpk ~]$pig

如果如下所示,则表示 Pig 已经正常启动。使用 quit 命令退出 Pig。

```
zkpk@master:~/pig-0.13.0/bin
File Edit View Search Terminal Help
2016-08-30 23:42:43,987 [main] INFO  org.apache.pig.impl.util.Utils - Default bo
otup file /home/zkpk/.pigbootup not found
2016-08-30 23:42:44,435 [main] INFO  org.apache.hadoop.conf.Configuration.deprec
ation - mapred.job.tracker is deprecated. Instead, use mapreduce.jobtracker.addr
ess
2016-08-30 23:42:44,452 [main] INFO  org.apache.hadoop.conf.Configuration.deprec
ation - fs.default.name is deprecated. Instead, use fs.defaultFS
2016-08-30 23:42:44,453 [main] INFO  org.apache.pig.backend.hadoop.executionengi
ne.HExecutionEngine - Connecting to hadoop file system at: hdfs://master:9000
SLF4J: Class path contains multiple SLF4J bindings.
SLF4J: Found binding in [jar:file:/home/zkpk/hadoop-2.5.1/share/hadoop/common/li
b/slf4j-log4j12-1.7.5.jar!/org/slf4j/impl/StaticLoggerBinder.class]
SLF4J: Found binding in [jar:file:/home/zkpk/hbase-0.98.7-hadoop2/lib/phoenix-4.
6.0-HBase-0.98-client.jar!/org/slf4j/impl/StaticLoggerBinder.class]
SLF4J: Found binding in [jar:file:/home/zkpk/hbase-0.98.7-hadoop2/lib/slf4j-log4
j12-1.6.4.jar!/org/slf4j/impl/StaticLoggerBinder.class]
SLF4J: See http://www.slf4j.org/codes.html#multiple_bindings for an explanation.
SLF4J: Actual binding is of type [org.slf4j.impl.Log4jLoggerFactory]
2016-08-30 23:42:44,865 [main] WARN  org.apache.hadoop.util.NativeCodeLoader - U
nable to load native-hadoop library for your platform... using builtin-java clas
ses where applicable
2016-08-30 23:42:45,573 [main] INFO  org.apache.hadoop.conf.Configuration.deprec
ation - fs.default.name is deprecated. Instead, use fs.defaultFS
grunt>
```

另外,Pig 具有许多用户可以使用的命令行选项。用户可以通过输入 pig -h 命令查看完整的选项列表。下面介绍几个常用的命令行。

(1) -h 或-help:列举可用的命令行选项,即 pig -h 或 pig -help。

(2) -e 或-execute:在 Pig 中单独执行一条命令,如 pig -e fs -ls,列出用户根目录下的文件。

(3) -h properties:列举出 Pig 将要使用的属性值。

(4) -P 或者-propertyFile:指定一个 Pig 应该读取的属性值配置文件。

(5) -version:打印出 pig 的版本信息。

本章小结

本章内容主要是为了简单地了解一下 Pig。

(1) 了解 Pig,并且深入了解 Pig 的设计思想。

(2) 介绍了如何完成 Pig 的安装部署,并且可以启动它。

习 题

1. 选择题

(1) 启动 Pig 的命令是(   )。

  A. start-all.sh B. stop-all.sh C. pig   D. 以上命令都可以

（2）关于 Pig，下列说法正确的是（ ）。

  A. Pig 的两种运行模式均需要 Hadoop 已经启动

  B. Pig local 模式不需要安装 JDK

  C. Pig 适合进行数据批处理

  D. 以上都不正确

2. 填空题

Pig 的两种运行模式分别为_____和_____。

3. 问答题

（1）Pig 是什么？它主要的职责是什么？

（2）简述 Pig 的安装和启动步骤。

第 7 章

Pig 基 础

本章摘要

Grunt 是运行 Pig 命令的交互式外壳环境。7.1 节主要介绍 Grunt 的主要用途、进入 Grunt 的方式、Grunt 和 HDFS 命令的交互使用，以及如何控制 Pig。

Pig Latin 是一种脚本语言。首先介绍 Pig Latin 语言的数据类型，包括 Pig 的基本数据类型、复杂数据类型、NULL 值和类型转换；然后以实例介绍 Pig Latin 的应用。

7.1 命令行工具 Grunt

Grunt 的主要用途是以交互式会话的方式输入 Pig Latin 脚本、作为访问 HDFS 的 shell 端口、提供了控制 Pig 和 Mapreduce 的命令。

7.1.1 输入 Pig Latin 脚本

如果实验环境资源足够多，建议将 Pig 安装在与 Hadoop 无关的节点上，把节点加入 Hadoop 集群网络中。因为 namenode 负责整个集群的管理，所以为了避免自己的错误操作导致集群崩溃，最好不要装在含 namenode 的节点上。如果集群不断运行 MapReduce 任务，DataNode 就会有较多的数据处理任务。

因为实验环境资源有限，这里只安装了 master 和 slave 节点，所以把 Pig 安装到了 master 节点上（参照第 6 章安装）。

Grunt 是运行 Pig 命令的交互式外壳环境，主要用途之一就是以交互式会话的方式输入 Pig Latin 脚本。Pig Latin 是一种脚本语言，Pig Latin 语言和传统关系数据库中的数据库操作语言（如 SQL 语言）非常类似，Pig Latin 语言侧重对数据的查询和分析，而且让用户每次只输入一条单独的语句，不像 SQL 一样要求用户一次性输入一条完成所有计算任务的语句。

不需要执行任何脚本和命令直接调用 Pig，就可以进入 Grunt shell。进入 Pig 的安装目录后，本地执行命令是 pig -x local，分布式执行命令是 pig -x mapreduce，与 HDFS 文件系统交互命令是 pig，立刻进入 Grunt shell 界面。用户可在 Grunt shell 命令中直接输入 Pig Latin 脚本语言。Pig 会对用户输入的 Pig Latin 脚本语言做一些基本的语法和语义检查，直到用户输入 store 或 dump 命令时，才执行 Pig Latin 脚本。

7.1.2 使用 HDFS 命令

Grunt 除了以交互式会话的方式输入 Pig Latin 脚本,Grunt 还是一个可以访问 HDFS 的 shell 端口。

输入 pig -x local,此时 pig 和本地的文件系统交互。

省略-x local,即直接在 Pig 安装目录下输入 pig,此时 pig 和 HDFS 交互。

注意:应确保 hadoop 已经正常启动。

1. 列出 HDFS 下的所有文件

```
grunt> fs -ls
Found 6 items
drwxr-xr-x   - zkpk supergroup          0 2016-07-31 20:56 .sparkStaging
drwxr-xr-x   - zkpk supergroup          0 2016-07-14 00:15 20111230
drwxr-xr-x   - zkpk supergroup          0 2016-07-11 09:58 QuasiMonteCarlo_14682
56281512_925922055
drwxr-xr-x   - zkpk supergroup          0 2016-09-10 20:54 output
drwxr-xr-x   - zkpk supergroup          0 2016-09-10 20:55 output1
drwxr-xr-x   - zkpk supergroup          0 2016-09-10 20:35 sogou
grunt> fs -lsr
```

该命令递归地显示子目录下的内容。

2. 列出 HDFS 文件下名为 output1 文件中的内容

```
grunt> fs -ls output1
Found 3 items
-rw-r--r--   1 zkpk supergroup          0 2016-07-20 22:37 output1/_SUCCESS
-rw-r--r--   1 zkpk supergroup          6 2016-07-20 22:37 output1/part-00000
drwxr-xr-x   - zkpk supergroup          0 2016-09-10 20:55 output1/sogou
```

3. 创建目录

```
grunt> fs -mkdir sogoudata
```

该命令只能一级一级地创建文件夹。

```
grunt> fs -ls
```

用该命令查看一下是否存在 sogoudata。若存在,则创建成功。

4. 上传文件

上传/home/zkpk 下的 sogou. demo 文件(截取的 sogou. 500w. utf8 前 100 条)搜狗数据的数据格式。

访问时间\t 用户 ID\t [查询词]\t 该 URL 在返回结果中的排名\t 用户点击的顺序号\t 用户点击的 URL

其中,用户 ID 是根据用户使用浏览器访问搜索引擎时的 Cookie 信息自动赋值,即同一次使用浏览器输入的不同查询对应同一个用户 ID。

将/home/zkpk 目录下的 sogou. demo 文件上传到 HDFS 上的 sogoudata 下。

```
grunt> fs -put /home/zkpk/sogou.demo sogoudata/
```

查看是否上传成功。

```
grunt> fs -ls sogoudata/
Found 1 items
-rw-r--r--   1 zkpk supergroup      10935 2016-10-07 22:32 sogoudata/sogou.demo
```

出现如上所示则证明上传成功。

-copyFromLocal 和-put 命令相同，其使用方法为：

```
fs -copyFromLocal 源路径 路径
```

把本地文件夹上传到 HDFS 文件夹 output1 中，代码如下：

```
grunt> fs -copyFromLocal /home/zkpk/sogoudata output1
```

查看 output1 是否已上传成功。

```
grunt> fs -ls output1/
Found 4 items
-rw-r--r--   1 zkpk supergroup          0 2016-07-20 22:37 output1/_SUCCESS
-rw-r--r--   1 zkpk supergroup          6 2016-07-20 22:37 output1/part-00000
drwxr-xr-x   - zkpk supergroup          0 2016-09-10 20:55 output1/sogou
drwxr-xr-x   - zkpk supergroup          0 2016-10-07 22:38 output1/sogoudata
```

5. 复制 HDFS 中的文件到本地系统中

将 HDFS 中的 sogou 文件复制到本地 /home/zkpk 下，并命名为 sogou。

```
grunt> fs -get sogou /home/zkpk/sogou
```

或

```
grunt> fs -copyToLocal sogou /home/zkpk/sogou
```

6. 删除文档

命令参数-rm 会删除一个文件或目录，而命令参数-rm -r 会删除一个文件或递归删除目录。

删除 HDFS 下名为 sogou 的文件。

```
grunt> fs -rm -r sogou
2016-10-07 22:44:59,156 [main] INFO  org.apache.hadoop.fs.TrashPolicyDefault - N
amenode trash configuration: Deletion interval = 0 minutes, Emptier interval = 0
 minutes.
Deleted sogou
```

7. 查看文件

查看 HDFS 下 sogoudata 文件中的内容。

```
grunt> fs -cat sogoudata/*
20111230000005	57375476989eea12893c0c3811607bcf	奇艺高清	1	1	http://www.qiyi.com/
20111230000005	66c5bb7774e31d0a22278249b26bc83a	凡人修仙传	3	1	http://www.booksky.org/BookDetail.aspx?BookID=1050804&Level=1
20111230000007	b97920521c78de70ac38e3713f524b50	本本联盟	1	1	http://www.bblianmeng.com/
20111230000008	6961d0c97fe93701fc9c0d861d096cd9	华南师范大学图书馆	1	1	http://lib.scnu.edu.cn/
20111230000008	f2f5a21c764aebde1e8afcc2871e086f	在线代理	2	1	http://proxyie.cn/
20111230000009	96994a0480e7e1edcaef67b20d8816b7	伟大导演	1	1	http://movie.douban.com/review/1128960/
20111230000009	698956eb07815439fe5f46e9a4503997	youku	1	1	http://www.youku.com/
20111230000009	599cd26984f72ee68b2b6ebefccf6aed	安徽合肥365房产网	1	1	http://hf.house365.com/
20111230000010	f577230df7b6c532837cd16ab731f874	哈萨克网址大全	1	1	http://www.kz321.com/
20111230000010	285f88780dd0659f5fc8acc7cc4949f2	IQ数码	1	1	http://www.iqshuma.com/
20111230000010	f4ba3f337efb1cc469fcd0b34feff9fb	推荐待机时间长的手机	1	1	http://mobile.zol.com.cn/148/1487938.html
```

8. 复制文件

复制文件通过命令参数-cp来完成，表示在 HDFS 中将 source 文件或目录复制到 destination。

```
fs -cp source destination
```

9. 通过 hdfs 命令把两个文件的内容合并

将 HDFS 中的多个源文件合并并保存到本地。

```
fs -getmerge  HDFS 中的源文件(里面有多个文件)合并后的文件名
```

例如，将 HDFS 上的 output1 目录下的文件合并后保存到本地 /home/zkpk/output1 文件夹下。

```
grunt>fs -getmerge output1 /home/zkpk/output1
```

10. 显示路径下所有文件磁盘使用情况

```
grunt> fs -du output1
0      output1/_SUCCESS
6      output1/part-00000
10935  output1/sogou
10935  output1/sogoudata
```

-du -s 显示全部文件或目录磁盘使用情况。

```
grunt> fs -du -s output1
21876   output1
```

此外，Grunt 本身也实现了一些命令，如 cd、pwd、cat、copyFromLocal、copyToLocal、cp、ls、mkdir、mv、rm、rmf 等。除了 cd、pwd 以外，其他都不推荐使用，而且它们在未来可能会被移除，所以推荐使用 HDFS 命令。

Grunt 增加了一个新命令 sh。该命令可以使用本地的 shell 命令，就像 FS 可以使用 HDFS 的命令一样。如 sh jps 查看当前进程。

7.1.3 控制 Pig

Grunt 提供了控制 Pig 和 Mapreduce 的命令 kill jobId。该命令可以终止指定 jobId 的 Mapreduce 任务。

以 7.1.2 小节中上传的文件 sogou.demo 为例（在 MapReduce 模式下运行，要保证 master 节点和 slave 节点都安装了 Pig，且 Hadoop 正常启动）说明 kill 命令的使用。

```
[zkpk@master pig-0.13.0]$pig -x mapreduce
grunt>sogoudata=load 'sogoudata/sogou.demo' as(ts,uid,keyword,rank,ord,url);
grunt>ordered=order sogoudata by rank;
grunt>subsogoudata=limit ordered 10;
grunt>DUMP subsogoudata;
```

注意：此代码暂时不需要理解。

当该任务运行时，会出现如下内容。

```
2016-10-08 00:58:24,384 [JobControl] INFO  org.apache.hadoop.mapreduce.Job - The
 url to track the job: http://master:18088/proxy/application_1475903930556_0005/
2016-10-08 00:58:24,384 [main] INFO  org.apache.pig.backend.hadoop.executionengi
ne.mapReduceLayer.MapReduceLauncher - HadoopJobId: job_1475903930556_0005
2016-10-08 00:58:24,384 [main] INFO  org.apache.pig.backend.hadoop.executionengi
ne.mapReduceLayer.MapReduceLauncher - Processing aliases ordered
```

可以看到该任务的 jobId 为 job_1475903930556_0005。

此时，以 MapReduce 模式下运行第二个 Grunt shell 窗口，确保在该任务还没有完成时，在第二个窗口中输入 kill job_1475903930556_0005，结束 MapReduce 任务。

```
grunt> kill job_1475903930556_0005
2016-10-08 00:58:41,052 [main] INFO  org.apache.hadoop.yarn.client.RMProxy - Con
necting to ResourceManager at master/192.168.111.128:18040
2016-10-08 00:58:48,654 [main] INFO  org.apache.pig.backend.hadoop.executionengi
ne.mapReduceLayer.MapReduceLauncher - Kill job 1475903930556_0005 submitted.
```

第一个 Grunt shell 窗口就会立即停止任务，并显示任务失败。

```
2016-10-08 00:58:49,777 [main] INFO  org.apache.pig.backend.hadoop.executionengi
ne.mapReduceLayer.MapReduceLauncher - Failed!
2016-10-08 00:58:49,782 [main] ERROR org.apache.pig.tools.grunt.Grunt - ERROR 10
66: Unable to open iterator for alias subsogoudata
Details at logfile: /home/zkpk/pig-0.13.0/pig_1475911913378.log
```

由上可知，kill jobId 会终止某个特定的任务。如果所终止的任务和其他任务无关，其他任务还会继续执行；但如果要终止其他任务，那么最好的方法是终止 Pig 进程，然后使用 kill 命令终止所有正在执行的 MapReduce 任务。

注意：终止 Pig 进程使用 Ctrl+C 组合键。

7.2 Pig 数据类型

在使用 Pig Latin 之前，需要知道该脚本语言的数据类型、出现缺失值的情况并了解如何处理 NULL 值，以及不同数据类型间的转换。Pig 的数据类型分为基本数据类型和复杂数据类型两大类。

7.2.1 基本类型

基本数据类型有 6 种，详情见表 7-1。基本数据类型在使用时只对应一个值。

表 7-1 基本数据类型

基本数据类型	描　　述	基本数据类型	描　　述
int	32 位有符号整数	double	64 位浮点数
long	64 位有符号整数	chararray	UTF-16 格式的字符数组
float	32 位浮点数	bytearray	字节数组

int、long、float、double 数据类型和 java 中对应的数据类型相同。chararray 类似于用 UTF-16 格式表示文本数据的 java.lang.String 字符串。bytearray 是一组字节，类似于表示二进制对象的 java 的 byte 数组，其是通过封装了 java 的 byte[] 的 DataByteArray java 类实现的。

7.2.2 复杂类型

复杂类型有 3 个,详情见表 7-2。复杂类型可以包含其他类型的数据。

表 7-2 复杂类型

复杂类型	描述
tuple	任何类型的字段序列
bag	元组的无序多重集合(允许重复元组)
map	一组键-值对。键是字符数组,值可以是任何类型的数据

1. tuple(元组)

一个 tuple 相当于 SQL 中的一行,tuple 字段相当于 SQL 中的列。tuple 数据类型是一组数据值的集合。tuple 可以分为多个字段,每个字段有序且对应着一个数据元素。数据元素可以是不同的数据类型。因为字段是有序的,所以可以在 tuple 中随机访问某个值。tuple 的格式如下:

(value1,value2,value3,...)

2. bag(包)

bag 数据类型是 tuple 和其他 bag 的容器。bag 是一个无序的 tuple 集合,所以不能随机访问 bag 中的某一个 tuple 或 bag。一个 bag 中允许出现重复的 tuple 或 bag。bag 的格式如下:

{(tuple1),(tuple2),...}

3. map(映射)

map 数据类型是关联数据类型,格式是由键-值对组成。可以通过键来查找对应数据元素的值。键是字符数据类型,即 chararray;值可以是任意类型的数据,默认类型是 bytearray。map 中的键-值对是唯一的。map 的格式如下:

[key1#value1,key2#value2,...]

7.2.3 NULL 值

定义与 SQL 中的 NULL 一样,即该值是未知的。任何数据类型的数据都可以是 NULL。产生 NULL 值的情况如下:

(1) 除数为 0。
(2) 一个值无法强制转换为声明的类型(7.2.4 小节详解)。
(3) 用户的自定义函数 UDF(8.3 节详解)。
(4) 引用一个不存在的字段。
(5) 引用一个 map 中不存在的字段。
(6) 引用一个 tuple 中不存在的字段。
(7) 加载不存在的数据时产生 NULL。空字符串不会被加载,会被替换成 NULL,NULL 可以作为一个常量使用。

(8) load 时数据类型不匹配产生 NULL。

综上所述，产生 NULL 值是因为处理数据时发生错误、数据缺失等原因造成的。

7.2.4 类型转换

用户想把不知道的数据类型转换成指定的数据类型时，可以对其进行数据转换。

各类型间的转换关系见表 7-3。

表 7-3 类型转换

from/to	bag	tuple	map	int	long	float	double	chararray	bytearray
bag		error	error	error	error	error	error	error	error
tuple	error		error	error	error	error	error	error	error
map	error	error		error	error	error	error	error	error
int	error	error	error		yes	yes	yes	yes	yes
long	error	error	error	yes		yes	yes	yes	error
float	error	error	error	yes	yes		yes	yes	error
double	error	error	error	yes	yes	yes		yes	error
chararray	error	error	error	yes	yes	yes	yes		error
bytearray	yes	yes	yes	yes	yes	yes	yes	yes	

由表 7-3 可知，转换成复杂数据类型或者将复杂数据类型转换为其他类型是不允许的，bytearray 类型转换为其他复杂数据类型除外。将 bytearray 类型转换为其他类型是允许的，但所有类型不允许转换为 bytearray。

下面举个简单的例子，以便了解 Pig 的基本类型、复杂类型、NULL 值、类型转换等。

在 /home/zkpk 下建立了一个 zkpk.txt 文件，里面的数据内容为：

```
[zkpk@master ~]$ cat zkpk.txt
r 1 2 3 4.1 5.9
r 3 3 4 3.6 2.6
r 1 2 3 5.2 3.6
r 1 5 6 7.8 1.3
r 2 4 8 9.2 2.3
n 1 5 6 - -
```

以上一共有 6 行数据，第 6 行最后两列有无效数据。

在本地模式下运行 Pig，使用命令 pig -x local。

(1) 根据第 2、3、4 列进行分类，然后求分类后的第 5、6 列的平均值。若有无效数据，则不用于平均值的计算。第 1、2 行为一条代码。

```
grunt> zkpk=load'/home/zkpk/zkpk.txt' using PigStorage(' ')as(col1:chararray,
col2:int,col3:int,col4:int,col5:double,col6:double);
grunt>A=GROUP zkpk BY(col2,col3,col4);
grunt>B=FOREACH A GENERATE group,AVG(zkpk.col5),AVG(zkpk.col6);
grunt>DUMP B;
```

输出结果如下：

```
((1,2,3),4.65,4.75)
((1,5,6),7.8,1.3)
((2,4,8),9.2,2.3)
((3,3,4),3.6,2.6)
```

(2) 统计数据的行数。

```
grunt> zkpk= load '/home/zkpk/zkpk.txt' using PigStorage(' ') as (col1:chararray,
col2:int,col3:int,col4:int,col5:double,col6:double);
grunt>A=group zkpk all;
grunt>B=foreach A gererate COUNT(zkpk);
grunt>DUMP B;
```

最终统计结果为 6。

建立 zkpk1.txt 文件，里面加入一计数列，内容如下：

```
[zkpk@master ~]$ cat zkpk1.txt
r 1 2 3 4.1 5.9 100
r 3 3 4 3.6 2.6 100
r 1 2 3 5.2 3.6 200
r 1 5 6 7.8 1.3 400
r 2 4 8 9.2 2.3 300
n 1 5 6 - - 400
```

(3) 在第 2、3、4 列的所有维度组合下，求最后一列不重复记录的条数。

```
grunt> zkpk= LOAD '/home/zkpk/zkpk1.txt' using PigStorage(' ')AS(col1:chararray,
col2:int, col3:int, col4:int, col5:double, col6:double, col7:int);
grunt>A=GROUP zkpk BY(col2, col3, col4);
grunt>B=FOREACH A {C=DISTINCT zkpk.col7; GENERATE group, COUNT(C);};
grunt>DUMP B;
```

输出结果如下：

```
((1,2,3),2)
((1,5,6),1)
((2,4,8),1)
((3,3,4),1)
```

(4) 类型转换。

在/home/zkpk 下建立了一个 test.txt 文件，里面的数据内容为：

```
[zkpk@master ~]$ cat test.txt
100,01
011,10
001,11
010,02
```

把第 1 列看成是字符数组，第 2 列是整型。

```
grunt>testdata=load '/home/zkpk/test.txt' using PigStorage(',')as(str:
chararray,number:int);
grunt>dump testdata;
```

输出结果如下：

```
(100,1)
(011,10)
(001,11)
(010,2)
```

把第 1 列 chararray 类型数据前 2 列数据强制转换为 int 类型。

```
grunt> result=foreach testdata generate(int)SUBSTRING(str,0,2);
grunt> dump result;
```

输出结果如下：

```
(10)
(1)
(0)
(1)
```

下面显示 chararray 类型转换成了 int 类型。

```
grunt> describe result;
result: {int}
```

把第 2 列 int 类型转换成 chararray 类型。

```
grunt> result1=foreach testdata generate(chararray)number;
grunt> dump result11;
grunt> describe result11;
```

下面显示 int 类型转换成了 chararray 类型。

```
grunt> describe result1;
result1: {number: chararray}
```

把第 2 列的 chararray 类型转换成 bytearray 类型时，报错，显示不能转换为 bytearray 类型。

```
grunt> result2 = foreach result1 generate(bytearray)number;
2016-10-08 19:46:58,704 [main] ERROR org.apache.pig.tools.grunt.Grunt - ERROR 10
51:
<line 5, column 35> Cannot cast to bytearray
Details at logfile: /home/zkpk/pig-0.13.0/pig 1475979587319.log
```

本章小结

本章内容主要是掌握 Grunt 的主要用途；掌握 Grunt 和 HDFS 交互使用时的命令并理解其含义；掌握 Pig 的数据类型：基本数据类型、复杂数据类型；了解 Pig 中产生 NULL 值的情况；掌握 Pig 中数据类型间的转换。

习 题

Pig 的数据类型有哪些？举例说明其复杂数据类型用于哪些场景？

第 8 章

Pig Latin 编程

本章摘要

第 6、7 章介绍了 Pig 的安装、命令行工具的使用以及数据类型,并通过简单的例子展示了 Pig 的基本命令,本章开始详细介绍 Pig 提供的 SQL-Like 编程语言 Pig Latin 的使用。

首先介绍 Pig Latin 相关的基本知识,然后给出 Pig Latin 提供的关系操作和相应的案例,接着将介绍用户自定义函数 UDF 的使用与编写,最后会介绍 Pig Latin 开发中的一些工具与技巧。

8.1 Pig Latin 介绍

8.1.1 基础知识

一个 Pig Latin 程序由一组语句构成,一个语句可以理解为一个操作或一个命令。为保证语句的正确性,一般以分号作为结束符。例如:

```
grunt>sogoudata=load 'sogoudata/sogou.demo' as(ts,uid,keyword,rank,ord,url);
grunt>ordered=order sogoudata by rank;
```

其中,sogoudata、ordered 是关系名称,关系名称即通常说的别名。一旦声明之后,那么这个分配是不变的。关系名称虽然可以被重复使用,但是建议不要这么做,因为这样做可能会丢失和之前重复的关系名称的联系。例 8-1 中 rank 是字段名称,字段名称代表的是一个关系所包含的字段的名称。它们的命名规则必须以字母字符开头,之后可以是字母、数字或下划线,其他所有字符必须都是 ASCII 码。

Pig Latin 有一个关键词列表,其中的单词在 Pig Latin 中有特殊含义,不能用于标识符。这些单词包括操作(load,illustrate)、命令(cat,ls)、表达式(matches,FLATTEN)和函数(DIFF,MAX)等。详见本章后面的介绍。

1. 大小写敏感性

Pig Latin 的大小写敏感性采用混合的规则。操作和命令是大小写无关的,如 load 和 group 这些命令;而关系名称、字段名称是大小写敏感的。用户自定义函数 UDF 的名称也是大小写敏感的。

2. 注释

Pig Latin 有两种注释方法。

(1) 双减号表示单行注释。Pig Latin 解析器会忽略从第一个减号开始到行尾的所有内容，例如：

```
grunt>--my program
grunt>DUMP A;
```

(2) C 语言风格的注释更灵活。这是因为它使用 /* 和 */ 符号表示注释开始和结束。这样，注释即可以跨多行，也可以内嵌在某一行内，例如：

```
Pig 代码示例
/*
 * description of my program
 * multiple lines.
 */
A=LOAD 'input/pig/join/A.txt';
Dump A;
```

8.1.2 输入和输出

在处理数据之前，应该知道如何为数据流增加输入和输出。

1. 加载

为了处理数据，首先需要使用 Pig 的 LOAD 函数从数据源加载数据，举个例子。

```
grunt>queries=LOAD "query_log.txt"
        USING myLoad()
        AS(userId, queryString, timestamp);
```

在上述例子中，LOAD 指定数据文件名（query_log.txt），这个路径可以是相对路径，可以是绝对路径，还可以指定一个完整的 URL 路径。数据通过 myLoad() 转换成一个个 tuple，每个 tuple 包含 userID、queryString、timestamp 三条数据信息。如果 USING 和 AS 省略，数据就会以默认的方式处理。

LOAD 处理完数据之后会生成一个包含 bag 的文件的 handle，并将其赋值给 queries 变量。之后就可以通过 queries 变量来处理数据了。

值得注意的是，LOAD 并没有像数据库那样将数据读进来，而是指定了数据该如何读取，只有在用户需要使用数据时才会被真正加载进来。

Pig 允许用户在加载数据时通过 using 句式指定加载函数，如果没有指定加载函数，默认使用内置的加载函数 PigStorage。以下代码是第 7 章讲解的例子，其中使用了加载函数，是想读取以空格分隔的 /home/zkpk/zkpk.txt 文件，as 句式可以为用户加载的数据指定字段和对应的数据类型。

```
grunt> zkpk= load '/home/zkpk/zkpk.txt' using PigStorage(' ') as (col1:chararray,
col2:int,col3:int,col4:int,col5:double,col6:double);
```

2. 存储

将数据加载到 Pig 中，并对数据进行一些有趣的操作后，就需要存储结果。Pig 提供了 store 语句进行写数据。

```
grunt> store subsogoudata into '/home/zkpk/sogou';
```

上面的语句，Pig 会将处理完的结果数据存储到/home/zkpk 路径下的 sogou 文件夹下，这个路径可以是相对路径，可以是绝对路径，还可以指定一个完整的 URL 路径。如果用户没有明确地指定存储函数，那么将会默认使用 PigStorage，用户也可以指定一个不同的存储函数。用户还可以给存储函数传参，例如，如果想将数据存储为以空格分隔的文本数据，PigStorage 会接受一个指定分隔符的参数。

```
grunt> store subsogoudata into '/home/zkpk/sogou' using PigStorage(' ')
```

3. 输出

有时用户不仅希望将结果存储到某个地方，还希望在控制台上看到结果数据。dump 语句可以将用户的脚本输出打印到屏幕上。

```
grunt> DUMP subsogoudata;
```

一直到 Pig 0.7 版本，dump 的输出数据符合 Pig Latin 中定义的常量格式。long 类型的值以 L 结尾，float 类型的值以 F 结尾，map 用[]分隔，tuple 使用()，bag 用{}。从 Pig 0.8 开始，long 值的 L 和 float 值的 F 被移除了，复杂类型的表示方式被保留下来了。输出中的每一条记录都是一个 tuple，所以是被()包围的，具体可参见 7.2.4 小节类型转换中列举的实例。

8.2 关系操作

Pig Latin 作为一种语言，关系型操作符是它最为显著的特征。下面将大部分关系操作列举在表 8-1 中。

表 8-1 Pig Latin 中的关系操作

指令	描述
load	加载数据
foreach	处理数据
filter	数据筛选
cogroup/group	在两个或更多关系中对数据归类/在一个关系中对数据进行分类
store/dump	存储数据结果/将结果打印到控制台上
join	连接
union	联合
cross	计算笛卡尔积
order	对数据进行排序
distinct	剔除重复项
limit	限制行数
sample	随机抽取
parallel	实现并行计算
dump	将数据打印到控制台
stream	使用外部程序对关系进行变换
split	把某个关系切分为两个或多个关系

关系操作符可以让用户对数据进行加载与存储、过滤、分组与连接、排序、合并与分割等转换。本节将介绍一些基本的关系操作符，以便用户能够使用 Pig Latin 编程。

8.2.1 foreach 语句

foreach 操作符主要用于处理数据，将输入关系的每一条记录转换成别的记录。foreach 语句接受的是一组表达式，在数据管道中将它们应用到每一条记录中，所以被命名为 foreach。通过该语句，会产生新的数据，并传送给下一个操作符。如下代码加载完整的记录，然后对每条记录只保留 uid、keyword 两个字段。

```
grunt>sogoudata=load'sogoudata/sogou.demo' as(ts,uid,keyword,rank,ord,url);
grunt>foreach sogoudata generate uid,keyword;
```

foreach 语句的表达式中最简单的就是常量和字段引用。字段引用可以通过别名引用或通过位置引用。位置引用是由 $ 符号和从 0 开始的整数组成。

```
grunt>zkpk=load'/home/zkpk/zkpk.txt' using PigStorage(' ') as (col1:chararray,
col2:int,col3:int,col4:int,col5:double,col6:double);
grunt>result1=foreach zkpk generate col4-col6;
grunt>result2=foreach zkpk generate $3-$5;
```

上面代码中，result1、result2 会存储相同的值。除了使用别名和位置，用户可以使用 * 代替所有字段。在版本 0.9 开始，用户也可以使用..（两个点）来指定字段区间，例如：

```
grunt>zkpk=load'/home/zkpk/zkpk.txt' using PigStorage(' ') as (col1:chararray,
col2:int,col3:int,col4:int,col5:double,col6:double);
grunt>begin=foreach zkpk generate ..col5;
grunt>mid=foreach zkpk generate clo2..col5;
grunt>end=foreach zkpk generate col5..;
```

begin 字段名称包括 col1 到 col5 字段，mid 字段名称包括 col2 到 col5 字段，end 字段名称包括 col5 和 col6 字段。Pig 对于基本的算术操作符是支持的，如加减乘除。Pig 也提供了三元条件操作符，这和 java 中的三元条件操作符类似，Pig 中如果前面的表达式返回的是 null 类型，则整个三元条件操作符的结果也为 null。为了从复杂类型中提取数据，要使用投射运算符。例如，map 使用 #，则后面跟着一个字符串类型的键的名称；tuple 是用 . 符号的，则后面跟着字段名；bag 的映射比较复杂，它不保证内部存放的 tuple 是以有序的方式存放的，因此对于一个 bag 中的 tuple 进行映射是没有意义的，当用户需要映射 bag 中的字段时，可以通过创建一个包含用户需要的字段的 bag。

8.2.2 filter 语句

filter 语句是做数据筛选的，通过 filter 语句用户可以选择哪些数据为之所用。filter 中包含了一个断言。在一条记录中，如果断言为 true，那么这条记录就可以在数据流中传下去，否则就不会。

断言表达式包含等值比较操作符，其中 == 和 != 可以用于复杂类型的比较。等值比较操作符都不可以用于 bag。对于 chararray 类型的数据，用户可以判断 chararray 是否符合指定的正则表达式。Pig 使用的是 Java 的正则表达式格式，例如，用户找所有包含字符串

abc 的字段,采用'.*abc.*'。如果想找不满足指定表达式的 chararray 数据时,可以在语句前面加上布尔操作符关键字 not。断言表达式也可以使用布尔操作符 and、or 连接多个断言表达式。例如:

```
grunt>sogoudata=load 'sogou/sogou.demo' using PigStorage('\t')as(ts:chararray,
uid:chararray,keyword:chararray,rank:int,ord:int,url:chararray);
grunt>aiqiyi=filter sogoudata by url matches '.*qiyi.*';
grunt>notaiqiyi=filter sogoudata by not url matches '.*qiyi.*';
```

aiqiyi 是选择 url 地址中有 qiyi 关键字的,notaiqiyi 是选择 url 地址中没有 qiyi 关键字的。

8.2.3 group 语句

group 语句和 SQL 中的语法相同,可以把具有相同键值的数据聚合在一起。Pig Latin 中的 group 操作与在 SQL 中的 group 操作还是有本质区别的。SQL 中的 group by 子句创建的组必须直接注入一个或多个聚合函数中,而 Pig 中的 group 和聚合函数之间没有直接的关系。它们处理 null 值的方式是一样的,即都是将以 null 作为键的所有记录汇集到相同的组里面。例如 7.2.4 小节中的例子。

```
grunt>zkpk=load '/home/zkpk/zkpk.txt' using PigStorage(' ')as(col1:chararray,
col2:int,col3:int,col4:int,col5:double,col6:double);
grunt>A=GROUP zkpk BY(col2,col3,col4);
grunt>B=FOREACH A GENERATE group,AVG(zkpk.col5),AVG(zkpk.col6);
```

上面的代码是以 col2、col3、col4 为键的值进行分组,然后求平均值。键值可以是一个或多个,也可以使用关键字 all 对用户的数据流中的所有字段进行分组。

group 操作符通常会触发一个 reduce 过程。分组是收集所有键中都包含相同值的记录。若数据流处于 map 阶段,那么就会迫使它先进行 shuffle,然后进行 reduce。若数据流处于 reduce 阶段,那么就会迫使它先进行 map,然后进行 shuffle,最后再进入 reduce 阶段。

8.2.4 order 语句

order 语句是对用户的数据进行排序,产生一个全排序的输出结果。全排序意味着不仅是将每个部分的数据进行排序,同时也会保证对于 n 个部分文件,也会保证第 n 个文件中的记录序号要比第 $n-1$ 个部分文件中的少。

order 语句的语法和 group 语句的语法类似。用户可以对一个或多个键进行排序。order 指定多个键时不需要使用圆括号。

```
grunt>sogoudata=load 'sogoudata/sogou.demo' as(ts,uid,keyword,rank,ord,url);
grunt>ordered=order sogoudata by rank;
```

上面实例是按照 rank 字段进行排序。默认是以升序排列的,当在指定键的后面加上 desc 关键字时,可以使结果降序排列。当 order 中指定多个键时,desc 关键字服从就近原则,仅对靠着它的那个键起作用。其他键还是按照默认的升序排列。

数据排序时,数值是按数字顺序排序,即所有数据类型中,null 值是最小的。chararray、

bytearray 的数据是按字典顺序排序,对复杂类型的数据进行排序会报错。

8.2.5 distinct 语句

distinct 语句会将重复值去掉,它只处理记录,不对字段级别进行计算。

```
grunt>--无重复总条数
grunt>sogoudata=load 'sogou/sogou.demo' using PigStorage('\t')as(ts:chararray,
uid:chararray,keyword:chararray,rank:int,ord:int,url:chararray);
grunt>classify=group sogoudata by(ts,uid,keyword,url);
grunt>result=DISTINCT classify;
```

它需要将相似的记录收集到一起,并判断它们是否重复。所有 distinct 会触发一个 reduce 处理过程。在 map 阶段可通过组合器将一些重复的数值去掉。

8.2.6 join 语句

join 是数据处理中非常重要的操作之一。它可以将一个输入中的记录和另一个输入中的数据结合在一起放置。把键值当作一个可以结合的条件,即键值相等时,数据才会被连接在一起,没有匹配的数据会被剔除。

```
grunt>sogoudata1=load 'sogou/sogou.demo' using PigStorage('\t')as(ts:chararray,
uid:chararray,keyword:chararray,rank:int);
grunt>sogoudata2= load 'sogou/sogou.txt' using PigStorage('\t')as(ts:chararray,
uid:chararray,keyword:chararray,url:chararray);
grunt>joinResult=join sogoudata1 by uid,sogoudata2 by uid;
```

用户可以指定一个或多个键,join 语句中关系名称都有相同数目的键,同时它们是相同的或者是同一类的数据类型。如果用户想在 join 操作之后引用字段 uid,需要使用 sogoudata1::uid 和 sogoudata2::uid 来加以区分。

Pig 也支持 outer join。在 outer join 中,没有和另一方匹配的数据也会被留下来,并使用 null 值填补缺失字段。外连接分为 left、right 和 full 3 种类型。左外连接即使在右边没有匹配的值,左边的数据也会被全部保留下来。右外连接即使在左边没有匹配的值,右边的数据也会被全部保留下来。全外连接即使两边没有匹配的值,两边的值也全部会保留下来。其中,关键词 outer 是可以省略的。

```
grunt>sogoudata1=load 'sogou/sogou.demo' using PigStorage('\t')as(ts:chararray,
uid:chararray,keyword:chararray,rank:int);
grunt>sogoudata2= load 'sogou/sogou.txt' using PigStorage('\t')as(ts:chararray,
uid:chararray,ord:int,url:chararray);
grunt>joinResult=join sogoudata1 by(ts,uid)left outer,sogoudata2 by(ts,uid);
```

8.2.7 limit 语句

使用 limit 操作符可以将关系减小到更易于操作的大小,可从结果中拿出几条数据查看。

```
grunt>sogoudata=load 'sogou/sogou.demo';
grunt>fist5=limit sogoudata 5;
```

在这个例子中将返回最多 5 条数据。除了 order 外的所有其他操作符，Pig 都不会保证产生的数据是按照一定次数排列的，因为给的数据是 100 条，所以上述脚本运行后可能返回不同的结果。但如果在 limit 前若加上 order 语句，就可保证每次执行返回结果是一样的。

8.2.8 sample 语句

sample 操作符将随机挑选关系元组的一个子集，用于抽取样本数据，需要表明采样关系中的记录的百分比。

```
grunt>logs=LOAD 'logs-test.txt' AS(id,key,ord);
grunt>sampled_logs=SAMPLE logs 0.15;
```

采样的量以所占关系中的元组的总体数量的百分比形式表示，范围从 0 到 1。在上面的示例中，0.15 表示提取了 15% 的数据量。使用 sample 脚本每次执行的结果都是不同的，百分比也并非是精确的，但是一定是近似的。

8.2.9 parallel 语句

Pig 的核心声明之一是它将提供一种并行数据处理语言。Pig 喜欢用户告诉其如何进行并行，所以 Pig 提供了 parallel 语句。

parallel 语句可以附加到 Pig Latin 中任一个关系操作符后。它只会控制 reduce 阶段的并行，因此只有对于可以触发 reduce 过程的操作符使用才有意义。在本地模式下 parallel 会被忽略，原因是本地模式下所有的操作符都是串行执行的。在下面例子中，parallel 会使 Pig 触发的 MapReduce 任务具有 10 个 reducer。

```
grunt>sogoudata=load 'sogou/sogou.demo' using PigStorage('\t')as(ts:chararray,
uid:chararray,keyword:chararray,rank:int,ord:int,url:chararray);
grunt>classify=group sogoudata by(ts,uid,keyword,url)parallel 10;
```

本节主要讲解 Pig 中的关系操作，下面举个例子让读者熟悉一下 Pig Latin 编程，还是使用 sogou 数据。保证 Hadoop 正常启动，使用 pig -x mapreduce 进入 grunt 命令行，使用上传到 HDFS 上的 sogou.demo 文件。

1. 从文件导入数据

```
grunt>sogoudata=load 'sogou/sogou.demo'  using PigStorage('\t')as(ts:chararray,
uid:chararray,keyword:chararray,rank:int,ord:int,url:chararray);
```

2. 查询整张表

```
grunt>DUMP gogoudata;
```

3. 查询 5 行数据

```
grunt>sogoudata_limit=limit sogoudata 5;
grunt>DUMP sogoudata_limit;
```

查询结果如下：

```
(20111230000005 57375476989eea12893c0c3811607bcf    奇艺高清                1    1
http://www.qiyi.com/,,,,,)
(20111230000005 66c5bb7774e31d0a22278249b26bc83a    凡人修仙传              3    1
http://www.booksky.org/BookDetail.aspx?BookID=1050804&Level=1,,,,,)
(20111230000007 b97920521c78de70ac38e3713f524b50    本本联盟                1    1
http://www.bblianmeng.com/,,,,,)
(20111230000008 6961d0c97fe93701fc9c0d861d096cd9    华南师范大学图书馆      1
1        http://lib.scnu.edu.cn/,,,,,)
(20111230000008 f2f5a21c764aebde1e8afcc2871e086f    在线代理                2    1
http://proxyie.cn/,,,,,)
```

4. 查询某些列

```
grunt>sogoudata_keyword=foreach sogoudata generate keyword;
grunt>DUMP sogoudata_keyword;
```

选取结果如下（截取部分结果）：

```
(奇艺高清)
(凡人修仙传)
(本本联盟)
(华南师范大学图书馆)
(在线代理)
(伟大导演)
(youku)
(安徽合肥365房产网)
(哈萨克网址大全)
(IQ数码)
(推荐待机时间长的手机)
(满江红)
(光标下载)
(张国立的电视剧)
(吹暖花开性吧)
```

5. 条数统计

（1）数据总条数。

```
grunt>cn=group sogoudata all;
grunt>count1=foreach cn generate COUNT(sogoudata);
grunt>DUMP count1;
2016-09-29 22:57:18,846 [main] INFO  org.apache.pig.backend.hadoop.executionengi
ne.mapReduceLayer.MapReduceLauncher - Success!
2016-09-29 22:57:18,854 [main] INFO  org.apache.hadoop.conf.Configuration.deprec
ation - fs.default.name is deprecated. Instead, use fs.defaultFS
2016-09-29 22:57:18,855 [main] INFO  org.apache.pig.data.SchemaTupleBackend - Ke
y [pig.schematuple] was not set... will not generate code.
2016-09-29 22:57:18,902 [main] INFO  org.apache.hadoop.mapreduce.lib.input.FileI
nputFormat - Total input paths to process : 1
2016-09-29 22:57:18,902 [main] INFO  org.apache.pig.backend.hadoop.executionengi
ne.util.MapRedUtil - Total input paths to process : 1
(100)
```

最终结果显示，共有 100 条数据。

（2）非空查询条数。

```
grunt>notNull=filter sogoudata by(keyword is not null and keyword !=' ');
grunt>cn=group notNull all;
grunt>count1=foreach cn generate COUNT(notNull);
grunt>DUMP count1;
```

（3）无重复总条数（根据 ts、uid、keyword、url）。

```
grunt>classify=group sogoudata by(ts,uid,keyword,url);
```

```
grunt>result=DISTINCT classify;
grunt>cn=group result all;
grunt>count1=foreach cn generate COUNT(result);
grunt>DUMP count1;
```

(4) 独立 UID 总数。

```
grunt>sogoudata=load 'sogou/sogou.demo' using PigStorage('\t')as(ts:chararray,
uid:chararray,keyword:chararray,rank:int,ord:int,url:chararray);
grunt>A=group sogoudata all;
grunt>B=FOREACH A {C=DISTINCT sogoudata.uid; GENERATE group, COUNT(C);}
grunt>DUMP B;
```

8.3 用户自定义函数 UDF

Pig 内置了许多函数供用户使用,大体有 4 种类型。

(1) 计算函数(Eval function):计算函数获取一个或多个表达式作为输入,并返回一个表达式。例如,MAX。

(2) 过滤函数(Filter function):过滤函数是一类特殊的计算函数。这类函数返回的是逻辑布尔值。例如,IsEmpty。

(3) 加载函数(Load function):加载函数指明如何从外部存储加载数据到一个关系。

(4) 存储函数(Store function):存储函数指明如何把一个关系中的内容存到外部存储。这些内置函数见表 8-2。

表 8-2 Pig 内置函数

类别	函数名称	描述
计算	AVG	计算包中项的平均值
	CONCAT	把两个字节数组或字符数组连接成一个
	COUNT	计算一个包中非空值的项的个数
	COUNTSTAR	计算一个包的项的个数,包括空值
	DIFF	计算两个包的差。如果两个参数不是包,那么如果它们相同,则返回一个包含这两个参数的包,否则返回一个空的包
	MAX	计算一个包中项的最大值
	MIN	计算一个包中项的最小值
	SIZE	计算一个类型的大小。数值型的大小总是1,对于字符数组,它返回字符的个数,对于字节数组,它返回字节的个数,对于容器(container,包括元组、包、映射),它返回其中项的个数
	SUM	计算一个包中项的值的总和
	TOKENIZE	对一个字符数组进行标记解析,并把结果词放入一个包
过滤	IsEmpty	判断一个包或映射是否为空

续表

类别	函数名称	描述
加载/存储	PigStorage	用字段分隔文本格式加载或存储关系。每一行被分为字段后[用一个可设置的分隔符（默认为一个制表符）分隔]，分别对应于元组的各个字段。这是不指定加载/存储方式时的默认存储函数
	BinStorage	从二进制文件加载一个关系或把关系存储到二进制文件中。该函数使用基于 Hadoop Writable 对象的 Pig 内部格式
	BinaryStorage	从二进制文件加载只包含一个类型为 bytearray 的字段的元组到关系，或以这种格式存储一个关系。bytearray 中的字节逐字存放。该函数与 Pig 的流式处理结合使用
	TestLoader	从纯文本格式加载一个关系。每一行对应于一个元组。每个元组只包含一个字段，即该行文本
	PigDump	用元组的 toString() 形式存储关系。每行一个元组，这个函数对设计很有帮助

如果表中没有需要的函数，可以自己写。Pig Latin 设计的基本理念是通过用户自定义函数（UDF）获得更好的扩展性，并为编写 UDF 提供一组良好定义的 API。Pig 的一大特色在于它允许用户通过 UDF 将 Pig、操作符和用户的代码或其他人提供的代码合并在一起使用。

8.3.1 注册 UDF

如果用户希望简化程序或重用程序代码，可以不使用 Pig 中的内置 UDF，这时可以使用自定义的函数。Pig 的用户自定义函数可以用 Java 编写，也可以用 Python 或 JavaScript 编写。

用户写 UDF 时要注意以下几点。

（1）UDF 传入的参数是 tuple。

（2）UDF 操作时，有点类似于对 group 后的结果进行操作，可以对 group 内的每一个元素进行操作，但是要记住每一个元素是什么，这样才能得到每一个元素。

（3）UDF 的使用。写好 UDF 后，打包 jar。如果用的是 Java，使用时只要像正常的 java 调用一样，把 class 路径写出来，把参数传进去就行了。

以第 7 章的 zkpk1.txt 文件为例。

```
[zkpk@master ~]$ cat zkpk1.txt
r 1 2 3 4.1 5.9 100
r 3 3 4 3.6 2.6 100
r 1 2 3 5.2 3.6 200
r 1 5 6 7.8 1.3 400
r 2 4 8 9.2 2.3 300
n 1 5 6 - - 400
grunt> zkpk=LOAD '/home/zkpk/zkpk1.txt' using PigStorage(' ') AS (col1:chararray,
col2:int, col3:int, col4:int, col5:double, col6:double, col7:int);
grunt> result=filter zkpk by col7!=100;
```

第二条语句的作用是筛选合法的数据。如果采用用户自定义函数，则第二条语句可以写为：

```
grunt>result=filter zkpk by IsBooleanExample(col7);
```

接下来,要自己定义一个 IsBooleanExample 函数,在 Java 中写成类,以 Java 代码为例。

```
package com.example.cn;
import java.io.IOException;
import org.apache.pig.FilterFunc;
import org.apache.pig.data.Tuple;

public class IsBooleanExample extends FilterFunc
{
    @Override
    Public Boolean exec(Tuple tuple)throws IOException{
        Object object=tuple.get(0);
        int col7=(Integer)object;
        return col7 !=100;
    }
}
```

此时需要完成如下操作。

(1) 编译代码,并打包成 jar 文件,如 UDF.jar。

(2) 通过 register 命令将这个 jar 文件注册到 pig 环境。参数为 jar 文件的本地路径。

```
grunt>register /home/zkpk/hadoop_jar/UDF.jar;
```

这样就可以使用自定义的函数。

```
grunt>result=filter zkpk by com.example.cn.IsBooleanExample(col7);
grunt>dump result;
```

但是,可以看到这个函数名太长,需要用定义别名的方式代替,以方便用户使用。使用 define 命令为其定义别名。

```
grunt>define IsBoolean com.example.cn.IsBooleanExample();
grunt>result=filter zkpk by IsBoolean(col7);
grunt>dump result;
```

总结自定义函数的实现方法,可发现:

(1) 需要定义一个继承自 FilterFunc 的类。

(2) 需要重写这个类的 exec 方法。这个方法的参数只有一个 tuple,但是调用时可以传递多个参数,可以通过索引号获得对应的参数值,如 tuple.get(0)表示取第一个参数。

(3) 调用时,需要使用类的全名(包括包名),所以常常会对其进行重定义别名。

(4) 更多的验证需要读者自行在函数中添加,如判断是否为 null 等。

8.3.2 define 命令和 UDF

正如前面提到的,define 命令可用于为用户的 Java UDF 定义一个别名,那么用户就不需要写那么冗长的包名全路径了。它也可以为用户的 UDF 的构造函数提供参数,亦可以用于定义 streaming 命令。

数学函数和过滤函数也是可以有一个或多个字符串类型的构造函数参数。如果用户使用的是一个接受构造函数参数的 UDF，define 命令后就可以放置这些参数，即参数传递时可以有一个参数，也可以有多个参数。

8.3.3 调用 Java 函数

Java 有丰富的工具集和函数库，因为 Pig 是使用 Java 实现的，所以 Java 中的一些函数可以供 Pig 用户使用。从 0.8 版本开始，Pig 提供了 invoker 方法，允许用户像使用 Pig UDF 一样使用一些静态 Java 函数。

所有的没有参数或有 int、long、float、double、String、array 类型的参数，同时有 int、long、float、double、String 返回值的静态函数都可以通过这种方式调用。

Pig Latin 不支持对返回值的数据类型进行重载，因此对于每一种类型有一个对应的调用方法：InvokerForInt、InvokerForLong、InvokerForFloat、InvokerForDouble、InvokerForString。用户需要根据期望的返回值的数据类型调用适当的调用方法。该方法有两个构造参数：第 1 个参数是完整的包名、类名、方法名；第 2 个参数是一个以空格作为分隔符的参数列表，这些参数将传送给这个 Java 函数，只包含参数的类型，这个参数可以是数组，也可以省略。对于 int、long、float、double，调用方法时参数是基本类型，但不可以是引用类型。例如，可以使用 int，但不能使用 Integer。

```
grunt>define hex
InvokeForString('java.lang.Integer.toHexString','int');
grunt> zkpk=LOAD '/home/zkpk/zkpk1.txt' using PigStorage(' ')AS(col1:chararray,
col2:int, col3:int, col4:int, col5:double, col6:double, col7:int);
grunt>IsNull=filter zkpk by col7 is not null;
grunt>tohex=foreach IsNull generate hex((int)col7);
```

8.4 开发工具

Pig 提供了一些工具来用于协助开发程序，本节将介绍一些调试工具，方便用户开发 Pig Latin 和提高开发效率。

8.4.1 describe

describe 命令会显示一个关系的模式（schema），当 Pig Latin 的使用者想要了解操作符是如何转换数据时，会发现这个命令很有用。下面通过例子介绍它的用法。

```
grunt>testdata=load '/home/zkpk/test.txt' using PigStorage(',')as(str:
chararray,number:int);
grunt>result1=foreach testdata generate(chararray)number;
grunt>dump result11;
grunt>describe result11;
```

describe 命令使用的是 Pig 的标准模式语法。该例子中的 test 文本文件包括 str 和 number 两个字段。其中，number 是 int 类型，然后把 number 字段从 int 类型转换成了 chararray 类型，如下结果显示 result1 包含一个字段 number，而且 number 已经成功转换为 chararray 类型。

```
grunt> describe result1;
result1: {number: chararray}
```

8.4.2　explain

　　explain 命令是用于洞察数据管道中 Pig 语句的执行计划的。当需要优化 Pig 脚本或调试错误时，它会很有帮助。它的输出包含三个部分：一个关系的逻辑计划、物理计划和 MapReduce 计划。具体来说，逻辑计划显示了管道和操作序列的应用；物理计划表明了该计划背景下的数据源和数据接收器；MapReduce 计划显示 MapReduce 在哪些地方需要应用操作符。

　　explain 是一个非常有用的衡量数据管道效率的操作。当用户试图优化脚本或者调试错误时，explain 命令显得特别有帮助。有两种方式使用 explain：用户可以在 explain 后加上 Pig Latin 中的任何一个别名；用户也可以在 grunt 命令行界面中对一个存在的脚本文件执行 explain。举一个执行脚本的例子，为简单起见，zkpk1.txt 里面的内容为：

```
[zkpk@master ~]$cat /home/zkpk/zkpk1.txt
1 100
3 100
1 200
1 400
2 300
1 400
grunt>zkpk= LOAD '/home/zkpk/zkpk1.txt' using PigStorage(' ')AS(col1:int,col2:int);
grunt>A=GROUP zkpk BY col1;
grunt>B=FOREACH A {C=DISTINCT zkpk.col2; GENERATE group, COUNT(C);};
grunt>store B into 'result';
```

　　该脚本文件命名为 changetype.pig。脚本文件中是根据第一列分组，然后在各组中统计第二列不同数出现的次数，执行以下命令：

```
[zkpk@master pig-0.13.0]$bin/pig -x local -e 'explain -script /home/zkpk/pig-0.13.0/bin/changetype.pig'
```

　　这将会产生一个文本格式的图像结构输出，可以随时检查输出信息。Pig 需要经历几个步骤才可以将一个 Pig Latin 脚本转换成一组 MapReduce 任务。进行基本的解析和语义检查之后，它将产生一个逻辑计划。该计划描述了 Pig 在执行这个脚本时将要使用的逻辑操作符，一些优化也会在此计划中完成。脚本 changetype.pig 生成的逻辑计划如图 8-1 所示。

　　这个图的流向是从底部到顶部的，所以 Load 操作符在最底部，连接线表示流向。从图 8-1 中可以看出脚本中的 4 个操作符（Load、Cogroup、Foreach、Store），操作符下垂的线对应的部分是附属于操作符的表达式。

　　对逻辑计划进行优化后，Pig 将会产生一个物理计划。物理计划描述了 Pig 在执行脚本时使用到的实体操作符，对应的物理计划如图 8-2 所示。

　　物理计划将要使用的实际路径和加载、存储函数已经确定了。该例子是在本地模式下进行的，所以该路径指的是本地文件。如果是在集群上执行，该路径会显示 HDFS 上的路径。

```
B: (Name: LOStore Schema: group#1:int,#16:long)
|
|---B: (Name: LOForEach Schema: group#1:int,#16:long)
    |   |
    |   (Name: LOGenerate[false,false] Schema: group#1:int,#16:long)ColumnPrune:
    OutputUids=[16, 1]ColumnPrune:InputUids=[1, 12]
    |   |   |
    |   |   group:(Name: Project Type: int Uid: 1 Input: 0 Column: (*))
    |   |   |
    |   |   (Name: UserFunc(org.apache.pig.builtin.COUNT) Type: long Uid: 16)
    |   |   |
    |   |   |---C:(Name: Project Type: bag Uid: 15 Input: 1 Column: (*))
    |   |
    |   |---(Name: LOInnerLoad[0] Schema: group#1:int)
    |   |
    |   |---C: (Name: LODistinct Schema: col2#2:int)
    |       |
    |       |---1-1: (Name: LOForEach Schema: col2#2:int)
    |           |   |
    |           |   (Name: LOGenerate[false] Schema: col2#2:int)
    |           |   |   |
    |           |   |   col2:(Name: Project Type: int Uid: 2 Input: 0 Column: (*
))
    |           |   |
    |           |   |---(Name: LOInnerLoad[1] Schema: col2#2:int)
    |           |
    |           |---zkpk: (Name: LOInnerLoad[1] Schema: col1#1:int,col2#2:int)
    |
    |---A: (Name: LOCogroup Schema: group#1:int,zkpk#12:bag{#19:tuple(col1#1:int
,col2#2:int)})
        |   |
        |   col1:(Name: Project Type: int Uid: 1 Input: 0 Column: 0)
        |
        |---zkpk: (Name: LOForEach Schema: col1#1:int,col2#2:int)
            |   |
            |   (Name: LOGenerate[false,false] Schema: col1#1:int,col2#2:int)Col
        umnPrune:OutputUids=[1, 2]ColumnPrune:InputUids=[1, 2]
            |   |   |
            |   |   (Name: Cast Type: int Uid: 1)
            |   |   |
            |   |   |---col1:(Name: Project Type: bytearray Uid: 1 Input: 0 Colu
mn: (*))
            |   |   |
            |   |   (Name: Cast Type: int Uid: 2)
            |   |   |
            |   |   |---col2:(Name: Project Type: bytearray Uid: 2 Input: 1 Colu
mn: (*))
            |   |
            |   |---(Name: LOInnerLoad[0] Schema: col1#1:bytearray)
            |   |
            |   |---(Name: LOInnerLoad[1] Schema: col2#2:bytearray)
            |
            |---zkpk: (Name: LOLoad Schema: col1#1:bytearray,col2#2:bytearray)Re
        quiredFields:null
#-----------------------------------------------
```

图 8-1　脚本 changetype.pig 生成的逻辑计划

```
#-----------------------------------------------
B: Store(file:///home/zkpk/pig-0.13.0/result:org.apache.pig.builtin.PigStorage)
- scope-22
|
|---B: New For Each(false,false)[bag] - scope-21
|   |
|   Project[int][0] - scope-12
|   |
|   POUserFunc(org.apache.pig.builtin.COUNT)[long] - scope-15
|   |
|   |---RelationToExpressionProject[bag][*] - scope-14
|       |
|       |---C: PODistinct[bag] - scope-20
|           |
|           |---1-1: New For Each(false)[bag] - scope-19
|               |   |
|               |   Project[int][1] - scope-17
|               |
|               |---Project[bag][1] - scope-16
|
|---A: Package(Packager)[tuple]{int} - scope-9
    |
    |---A: Global Rearrange[tuple] - scope-8
        |
        |---A: Local Rearrange[tuple]{int}(false) - scope-10
            |   |
            |   Project[int][0] - scope-11
            |
            |---zkpk: New For Each(false,false)[bag] - scope-7
                |   |
                |   Cast[int] - scope-2
                |   |
                |   |---Project[bytearray][0] - scope-1
                |   |
                |   Cast[int] - scope-5
                |   |
                |   |---Project[bytearray][1] - scope-4
                |
                |---zkpk: Load(/home/zkpk/zkpk1.txt:PigStorage(' ')) - scope
-0
```

图 8-2 物理计划

Pig 完成该计划后，Pig 会有一个 MapReduce 计划。该计划描述 map 过程、组合过程和 reduce 过程，如图 8-3 是 MapReduce 计划，这个和物理计划很相似。数据流在这里分解成了 3 个执行过程：Map 过程、Combine 过程和 Reduce 过程。这里没有 Global Rearrange 操作符了，因为它是 shuffle 阶段的一个替代。如果这个例子有多个 MapReduce 任务，那么它们都会在这个输出中展现出来。

8.4.3 illustrate

illustrate 操作符实际检查管道并生成数据，以确保管道中的每个环节都有数据通过，这样可以在处理真实的大规模数据之前，使用一个小的数据集测试整个管道，从而节省大量的调试时间。

与使用 describe 一样，如果想使用 illustrate，只需要在相应的别名之前使用即可。还是以前面的脚本为例。

```
[zkpk@master pig-0.13.0]$bin/pig -x local -e 'illustrate -script /home/zkpk/pig-
0.13.0/bin/changetype.pig'
```

```
MapReduce node scope-23
Map Plan
A: Local Rearrange[tuple]{int}(false) - scope-39
|   |
|   Project[int][0] - scope-40
|
|---B: New For Each(false,false)[bag] - scope-25
    |   |
    |   Project[int][0] - scope-26
    |   |
    |   POUserFunc(org.apache.pig.builtin.Distinct$Initial)[tuple] - scope-27
    |   |
    |   |---1-1: New For Each(false)[tuple] - scope-29
    |       |   |
    |       |   Project[int][1] - scope-28
    |       |
    |       |---Project[bag][1] - scope-30
    |
    |---Pre Combiner Local Rearrange[tuple]{Unknown} - scope-41
        |
        |---zkpk: New For Each(false,false)[bag] - scope-7
            |   |
            |   Cast[int] - scope-2
            |   |
            |   |---Project[bytearray][0] - scope-1
            |   |
            |   Cast[int] - scope-5
            |   |
            |   |---Project[bytearray][1] - scope-4
            |
            |---zkpk: Load(/home/zkpk/zkpk1.txt:PigStorage(' ')) - scope-0------
--
Combine Plan
A: Local Rearrange[tuple]{int}(false) - scope-43
|   |
|   Project[int][0] - scope-44
|
|---B: New For Each(false,false)[bag] - scope-31
    |   |
    |   Project[int][0] - scope-32
    |   |
    |   POUserFunc(org.apache.pig.builtin.Distinct$Intermediate)[tuple] - scope-33
    |   |
    |   |---Project[bag][1] - scope-34
    |
    |---A: Package(CombinerPackager)[tuple]{int} - scope-37--------
Reduce Plan
B: Store(file:///home/zkpk/pig-0.13.0/result:org.apache.pig.builtin.PigStorage)
 - scope-22
|
|---B: New For Each(false,false)[bag] - scope-21
    |   |
    |   Project[int][0] - scope-12
    |   |
    |   POUserFunc(org.apache.pig.builtin.COUNT)[long] - scope-15
    |   |
    |   |---POUserFunc(org.apache.pig.builtin.Distinct$Final)[bag] - scope-24
    |       |
    |       |---Project[bag][1] - scope-35
    |
    |---A: Package(CombinerPackager)[tuple]{int} - scope-9--------
Global sort: false
----------------
```

图 8-3 MapReduce 计划

  illustrate 是在 Pig 0.2 版本中加入的，但是有段时间并没有对它进行很好的维护。在 0.9 版本中，对它进行了重构。在 0.7 和 0.8 版本中，它对一些 Pig 操作符是有作用的，但并非支持所有的操作符。下面是该命令的输出结果。

```
---------------------------------------------
| zkpk       | col1:int    | col2:int        |
---------------------------------------------
|            | 1           | 400             |
|            | 1           | 200             |
---------------------------------------------

---------------------------------------------------------------
| A          | group:int   | zkpk:bag{:tuple(col1:int,col2:int)} |
---------------------------------------------------------------
|            | 1           | {(1, 400), (1, 200)}              |
---------------------------------------------------------------

-------------------------------
| B.1 1      | col2:int      |
-------------------------------
|            | 400           |
-------------------------------

-------------------------------
| B.C        | col2:int      |
-------------------------------
|            | 400           |
-------------------------------

-----------------------------------------
| B          | group:int   | :long       |
-----------------------------------------
|            | 1           | 2           |
-----------------------------------------

-----------------------------------------
| Store : B  | group:int   | :long       |
-----------------------------------------
|            | 1           | 2           |
-----------------------------------------
```

8.4.4 Pig 统计信息

从 0.8 版本开始,在每次执行结束后,Pig 都会产生一组统计信息的总结。

```
grunt> zkpk= LOAD '/home/zkpk/zkpk1.txt' using PigStorage(' ')AS(col1:int,col2:int);
grunt>A=GROUP zkpk BY col1;
grunt>B=FOREACH A {C=DISTINCT zkpk.col2; GENERATE group, COUNT(C);};
grunt>dump B;
```

执行这个代码后,会出现如下信息。

```
HadoopVersion   PigVersion      UserId    StartedAt            FinishedAt           Features
2.5.1           zkpk            2016-10-23 05:48:48   2016-10-23 05:48:50  GROUP_BY

Success!

Job Stats (time in seconds):
JobId    Maps   Reduces MaxMapTime    MinMapTIme    AvgMapTime    MedianMa
lias    Feature Outputs
job_local100103557_0002 1       1       n/a    n/a    n/a    n/a
p25371855/tmp-837398756,

Input(s):
Successfully read 6 records from: "/home/zkpk/zkpk1.txt"

Output(s):
Successfully stored 3 records in: "file:/tmp/temp25371855/tmp-837398756"

Counters:
Total records written : 3
Total bytes written : 0
Spillable Memory Manager spill count : 0
Total bags proactively spilled: 0
Total records proactively spilled: 0
```

```
Job DAG:
job_local100103557_0002


2016-10-23 05:48:50,383 [main] INFO  org.apache.pig.backend.hadoop.executionengi
2016-10-23 05:48:50,389 [main] INFO  org.apache.hadoop.conf.Configuration.deprec
sum
2016-10-23 05:48:50,396 [main] INFO  org.apache.hadoop.conf.Configuration.deprec
2016-10-23 05:48:50,397 [main] WARN  org.apache.pig.data.SchemaTupleBackend - Sc
2016-10-23 05:48:50,462 [main] INFO  org.apache.hadoop.mapreduce.lib.input.FileI
2016-10-23 05:48:50,462 [main] INFO  org.apache.pig.backend.hadoop.executionengi
(1,3)
(2,1)
(3,1)
```

前3行给出了这个任务的一个简要的总结。StartedAt 是 Pig 提交这个任务的开始时间，而不是第一个任务开始在 Hadoop 集群上开始执行的时间，取决于用户集群的繁忙程度，这个值可能有很大的差异。FinishedAt 是 Pig 完成处理这个任务的结束时间，这个时间会比最后一个 MapReduce 任务结束的时间稍微晚些。

Job Stats 标记的第二节提供了一个对每个执行的 MapReduce 任务的分解统计信息，统计信息包含每个任务具有的 map 个数和 reduce 个数，以及这些任务消耗的时间，还有用户的 Pig Latin 脚本和这些任务对应的别名信息。这个显示对应别名的功能是非常重要的，特别是用户想知道他的脚本的哪个操作符在哪个 MapReduce 任务中执行，这对于判断为什么一个特定的任务失败了或者产生了非预期的结果是有帮助的。上面是在本地模式下运行的，所以没有显示。以相同的例子在 MapReduce 模式下运行的结果如下：

```
HadoopVersion   PigVersion      UserId   StartedAt              FinishedAt             Features
2.5.1           zkpk            2016-10-23 06:18:14    2016-10-23 06:18:54    GROUP_BY

Success!

Job Stats (time in seconds):
JobId           Maps     Reduces MaxMapTime      MinMapTIme      AvgMapTime      MedianMa
pTime   MaxReduceTime   MinReduceTime   AvgReduceTime   MedianReducetime        A
lias    Feature Outputs
job_1477227613599_0002 1        1       6       6       6       6
6       6       1-124,A,B,zkpk  GROUP_BY,COMBINER       hdfs://master:9000/tmp/t
emp-939969037/tmp-235762810,

Input(s):
Successfully read 6 records (393 bytes) from: "hdfs://master:9000/user/zkpk/zkpk
1.txt"

Output(s):
Successfully stored 3 records (21 bytes) in: "hdfs://master:9000/tmp/temp-939969
037/tmp-235762810"

Counters:
Total records written : 3
Total bytes written : 21
Spillable Memory Manager spill count : 0
Total bags proactively spilled: 0
Total records proactively spilled: 0

Job DAG:
job_1477227613599_0002
```

```
2016-10-23 06:18:54,812 [main] INFO  org.apache.pig.backend.hadoop.executionengi
ne.mapReduceLayer.MapReduceLauncher - Success!
2016-10-23 06:18:54,815 [main] INFO  org.apache.pig.data.SchemaTupleBackend - Ke
y [pig.schematuple] was not set... will not generate code.
2016-10-23 06:18:54,826 [main] INFO  org.apache.hadoop.mapreduce.lib.input.FileI
nputFormat - Total input paths to process : 1
2016-10-23 06:18:54,826 [main] INFO  org.apache.pig.backend.hadoop.executionengi
ne.util.MapRedUtil - Total input paths to process : 1
(1,3)
(2,1)
(3,1)
```

Input、Output、Counters 是 Pig 为了避免耗尽内存资源对数据进行分隔的次数等相关的统计信息。从路径可以看出哪个在本地模式下运行，哪个在 MapReduce 模式下运行。在 Job DAG 中可以看出 MapReduce 任务之间数据是如何流向的，在该例子中只开启了一个任务。

8.4.5　M/R 作业状态信息

当用户在 Hadoop 集群上执行 Pig Latin 脚本时，找到作业的状态和日志信息是很有挑战性的。

定位到日志的第一步就是访问 Yarn 应用程序的管理网页。网址是 http://master：18088。其中，master 是用户在配置文件中配置的监控 Job 资源调度的域名。如图 8-4 所示为其管理页面。

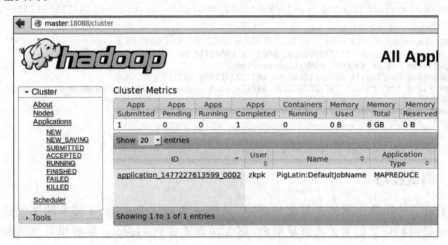

图 8-4　管理界面

图 8-4 只截取了一部分，由这张截图可以看到，在伪分布式集群上最近只运行过一个作业，在页面中用户可以看到是谁提交了作业、作业名称、作业类型等。8.4.4 小节中 Job Stats 有一个 JobId，在 Yarn 管理页面中的 ID 也会出现，可以发现这个页面对应的 ID 和 JobId 相同。当这些 M/R 作业真正在集群上运行时，才会显示在 Yarn 管理页面上。

单击作业 ID 后，显示如图 8-5 所示界面。

这个页面显示该作业的执行信息，包括作业的开始时间、状态、结束状态等信息。当想看该作业的日志信息时，可以点击 logs 查看日志，如图 8-6 所示。

也可以进入从节点查看。

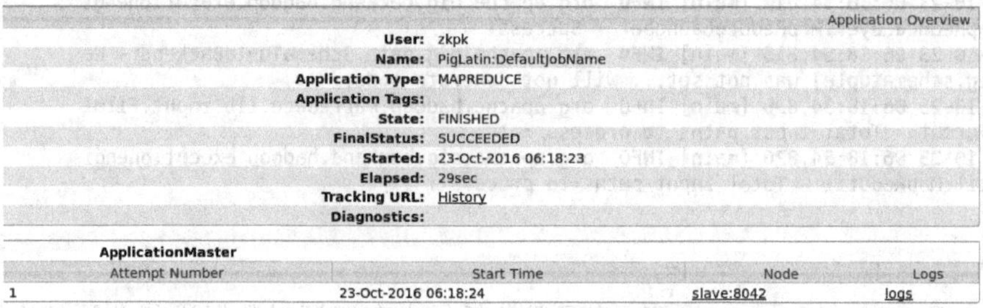

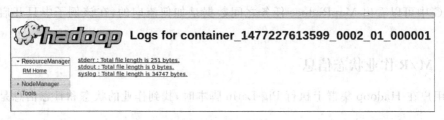

图 8-5 作业的执行信息

图 8-6 作业的日志信息

```
[zkpk@slave ~]$ cd $HADOOP_HOME/logs/userlogs/
[zkpk@slave userlogs]$ cd application_1477227613599_0002/
[zkpk@slave application_1477227613599_0002]$ cd container_1477227613599_0002_01
_000001
[zkpk@master container_1477227613599_0002_01_000001]$ ls
[zkpk@slave container_1477227613599_0002_01_000001]$ cd
[zkpk@slave ~]$ cd $HADOOP_HOME/logs/userlogs/
[zkpk@slave userlogs]$ cd application_1477227613599_0002/
[zkpk@slave application_1477227613599_0002]$ cd container_1477227613599_0002_01_
000001
[zkpk@slave container_1477227613599_0002_01_000001]$ ls
stderr  stdout  syslog
```

然后使用 cat 命令查看日志信息。

8.4.6 调试技巧

除了前 5 个小节介绍的工具之外，还有其他一些调试 Pig Latin 脚本的注意点。如果 illustrate 提供的信息仍然无法满足用户，那么在真正放到 Hadoop 集群执行之前使用本地模式对脚本进行测试。

使用本地模式的优点如下。

（1）尽管还是慢，但是会比使用 Hadoop 网格要快，因为后者需要等待资源槽位，而 job 初始化过程一般都要至少 30 秒的时间。

（2）所有操作符产生的日志从屏幕上都可以看到，而不是放在某个任务节点上。

（3）本地模式下使用的都是本地进程，这意味着用户可以对这个进程进行调试，这在调试 UDF 时特别有帮助。

另外，可以通过关闭一些特定的功能来看是否是因为这些功能造成了问题。表 8-3 列举出可以关闭的功能，这些功能都是可以通过命令行参数传递给 Pig 的。

表 8-3 可以关闭的功能列表

命令行选项	命令作用	什么情况需要关闭
-t SplitFilter	阻止 Pig 对过滤器谓词进行分隔，防止部分谓词前推到数据流上端	过滤器移除的行并非是预期的
-t MergeFilter	阻止 Pig 为了更高效的计算而合并邻近的操作符	过滤器移除的行并非是预期的
-t PushUpFilter	阻止 Pig 将数据流中过滤器操作符前置到邻近的操作符之前	过滤器移除的行并非是预期的
-t PushDownForEachFlatten	阻止 Pig 将数据流中包含 flatten 操作的 foreach 操作符前置到邻近的其他操作符之前	foreach 并没有产生预期的行或者字段
-t ColumnMapKeyPrune	阻止 Pig 来预测脚本中真正使用的那些字段，并让加载器只加载这些字段	加载函数返回的字符并非是预期的
-t LimitOptimizer	阻止 Pig 在数据流中将 limit 操作符前置到邻近的其他操作符之前	limit 操作并没有返回预期的指定行数
-t AddForEach	阻止 Pig 在脚本中加入 foreach 操作符，以移除不需要的字段	结果中没有期望的字段
-t MergeForEach	阻止 Pig 为了提高运算效率合并邻近的 foreach 操作符	foreach 并没有产生预期的行或者列
-t LogicalExpressionsSimplifier	阻止 Pig 进行一些表达式简化优化	foreach 并没有产生预期的行或者列
-t All	关闭所有逻辑优化策略。物理优化策略（例如使用组合器、multiquery 等）依旧是生效的	脚本并没有产生预期的行，同时需要判断逻辑优化器是否是产生问题的原因之一

本章小结

对于程序员、数据分析师和数据科学家而言，Pig 是一种相当有用的工具，它对 MapReduce 进行了一个更高层次的抽象，而 Pig Latin 是实现其分析功能的重要编程语言。本章介绍了有用的加载、处理和存储数据的技术，使用各种关系运算符完成基本操作，并且进一步介绍了用户如何通过自定义函数来实现更复杂的功能，最后对开发过程中常用的工具进行了介绍，这些工具可以通过洞察整个 Pig 的执行来帮助 Pig 使用者更好地完成工作。

习　题

1. 填空题

（1）如果用户需要在控制台上看到结果数据 resultData，则相应的操作语句为_____。

（2）列出以下关系操作符的含义：filter _____，distinct _____，sample _____，parallel _____。

(3) Pig 内置的函数大致包含四种类型：_____、_____、_____、_____。如果用户使用的功能不包含在这四类中,用户还可以通过_____来实现功能。

2. 操作题

(1) 查找最高气温：利用 Pig 统计每年的最高气温。数据文件内容如下(每行一个记录,tab 分割)。

```
1990 21
1990 18
1991 21
1992 30
1992 999
1990 23
```

并思考,如果采用用户自定义函数完成判定温度是否合法的操作？如何实现？

(2) 数据去重：按照指定字段进行数据去重。数据如下所示,请按照 1、2、3、4 字段进行去重。

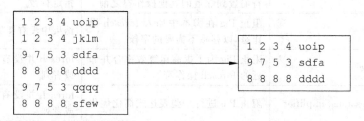

第 9 章

数据 ETL 工具 Sqoop

本章摘要

Sqoop 可以在 Hadoop 和关系型数据库之间转移大量数据。Sqoop 项目开始于 2009 年,最早是作为 Hadoop 的一个第三方模块存在,后来为了让使用者能够快速部署,也为了让开发人员能够更快速地迭代开发,Sqoop 独立成为一个 Apache 项目。

Sqoop 类似于其他 ETL 工具,使用元数据模型来判断数据类型,并确保从数据源转移到 Hadoop 时数据处理的类型安全。Sqoop 专为大数据批量传输设计,能够分割数据集,并创建 Hadoop 任务来处理每个区块。

本章首先在 Hadoop 集群上安装部署 Sqoop,然后通过实例来了解 Sqoop 的工作原理。

9.1 安装 Sqoop

Sqoop 的安装需要在 Hadoop 已经成功安装的基础上,并且要求 Hadoop 已经正常启动。下面的操作都假设通过 HadoopMaster 节点,所有的操作都使用 zkpk 用户。切换用户的命令是:

[zkpk@master ~]$ su zkpk

密码是:

zkpk

1. 解压并安装 Sqoop

使用下面的命令,解压 Sqoop 安装包。

[zkpk@master ~]$ cd /home/zkpk/software/hadoop/apache
[zkpk@master apache]$ mv sqoop-1.4.5.bin_hadoop-2.0.4-alpha.tar.gz ~/
[zkpk@master apache]$ cd
[zkpk@master ~]$ tar -zxvf sqoop-1.4.5.bin_hadoop-2.0.4-alpha.tar.gz
[zkpk@master ~]$ cd sqoop-1.4.5.bin_hadoop-2.0.4-alpha

执行下面的 ls -l 命令,会看到 Sqoop 包含的文件。

```
[zkpk@master sqoop-1.4.5.bin_hadoop-2.0.4-alpha]$ ls -l
total 1712
drwxr-xr-x. 2 zkpk zkpk  4096 Apr 12 22:41 bin
-rw-rw-r--. 1 zkpk zkpk 58531 Aug  1  2014 build.xml
-rw-rw-r--. 1 zkpk zkpk 29159 Aug  1  2014 CHANGELOG.txt
-rw-rw-r--. 1 zkpk zkpk  9273 Aug  1  2014 COMPILING.txt
drwxr-xr-x. 2 zkpk zkpk  4096 Apr 12 22:41 conf
```

```
drwxr-xr-x. 5 zkpk zkpk   4096 Apr 12 22:41 docs
drwxr-xr-x. 2 zkpk zkpk   4096 Apr 12 22:41 ivy
-rw-rw-r--. 1 zkpk zkpk  16465 Aug  1  2014 ivy.xml
drwxr-xr-x. 2 zkpk zkpk   4096 Apr 12 22:41 lib
-rw-rw-r--. 1 zkpk zkpk  19796 Aug  1  2014 LICENSE.txt
-rw-rw-r--. 1 zkpk zkpk    256 Aug  1  2014 NOTICE.txt
-rw-rw-r--. 1 zkpk zkpk  18772 Aug  1  2014 pom-old.xml
-rw-rw-r--. 1 zkpk zkpk   1096 Aug  1  2014 README.txt
-rw-rw-r--. 1 zkpk zkpk 967124 Aug  1  2014 sqoop-1.4.5.jar
-rw-rw-r--. 1 zkpk zkpk 574152 Aug  1  2014 sqoop-test-1.4.5.jar
drwxr-xr-x. 8 zkpk zkpk   4096 Aug  1  2014 src
drwxr-xr-x. 4 zkpk zkpk   4096 Apr 12 22:41 testdata
```

2. 配置 Sqoop

(1) 配置 MySQL 连接器。将 MySQL 的 java connector 复制到依赖库中。

```
[zkpk@master ~]$ cd /home/zkpk/software/mysql
[zkpk@master mysql]$ tar -zxvf mysql-connector-java-5.1.27.tar.gz
[zkpk@master mysql]$ mv mysql-connector-java-5.1.27/mysql-connector-java-5.1.27
-bin.jar ~/sqoop-1.4.5.bin_hadoop-2.0.4-alpha/lib/
```

(2) 配置环境变量。利用模板，准备配置文件。

```
[zkpk@master ~]$ cd ~/sqoop-1.4.5.bin_hadoop-2.0.4-alpha/conf
[zkpk@master conf]$ cp sqoop-env-template.sh sqoop-env.sh
```

将该文件 sqoop-env.sh 内容修改为（各个组件的安装路径请配置成用户自身的路径）：

```
#See the License for the specific language governing permissions and
#limitations under the License.

#included in all the hadoop scripts with source command
#should not be executable directly
#also should not be passed any arguments, since we need original $ *

#Set Hadoop-specific environment variables here.

#Set path to where bin/hadoop is available
#export HADOOP_COMMON_HOME=/home/zkpk/hadoop-2.5.2

#Set path to where hadoop-*-core.jar is available
#export HADOOP_MAPRED_HOME=/home/zkpk/hadoop-2.5.2

#set the path to where bin/hbase is available
#export HBASE_HOME=/home/zkpk/hbase-0.98.7-hadoop2

#Set the path to where bin/hive is available
#export HIVE_HOME=/home/zkpk/apache-hive-2.1.1-bin

#Set the path for where zookeper config dir is
#export ZOOCFGDIR=
```

3. 启动并验证 Sqoop

进入 Sqoop 安装主目录。

```
[zkpk@master ~]$cd ~/sqoop-1.4.5.bin_hadoop-2.0.4-alpha
[zkpk@master sqoop-1.4.5.bin_hadoop-2.0.4-alpha]$bin/sqoop help
```

执行命令后,会看到下面的打印输出,表示安装成功。

```
Please set $ACCUMULO_HOME to the root of your Accumulo installation.
Warning: /home/zkpk/sqoop-1.4.5.bin_hadoop-2.0.4-alpha/bin/../../zookeeper does
    not exist! Accumulo imports will fail.
Please set $ZOOKEEPER_HOME to the root of your Zookeeper installation.
16/04/12 23:23:44 INFO sqoop.Sqoop: Running Sqoop version: 1.4.5
usage: sqoop COMMAND [ARGS]

Available commands:
  codegen            Generate code to interact with database records
  create-hive-table  Import a table definition into Hive
  eval               Evaluate a SQL statement and display the results
  export             Export an HDFS directory to a database table
  help               List available commands
  import             Import a table from a database to HDFS
  import-all-tables  Import tables from a database to HDFS
  job                Work with saved jobs
  list-databases     List available databases on a server
  list-tables        List available tables in a database
  merge              Merge results of incremental imports
  metastore          Run a standalone Sqoop metastore
  version            Display version information

See 'sqoop help COMMAND' for information on a specific command.
```

9.2 数据导入

Sqoop 是一款 Apache 开源的工具,主要用于在 Hadoop 的 Hive 与传统的数据库间传递数据。可以将一个关系型数据库中的数据导入到 Hadoop 的文件系统 HDFS 中,也可以将 HDFS 的数据导出到关系型数据库中。图 9-1 所示为 Sqoop 的导入过程。

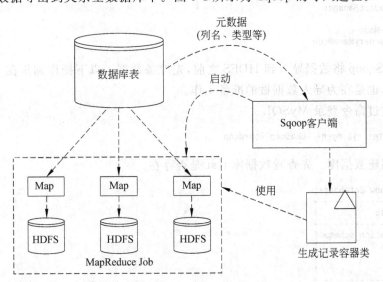

图 9-1 Sqoop 的导入过程

图 9-1 演示了 Sqoop 是如何与关系数据库和 Hadoop 进行交互的。Sqoop 像 Hadoop 一样,也是用 Java 语言编写的。Java 提供了一个称为 JDBC(Java Database Connectivity)的

API，应用程序可以使用这个 API 来访问存储在 RDBMS 中的数据，而且同时可以检查数据类型。

在导入数据之前，Sqoop 使用 JDBC 来检查将要导入的表。它检索出表中所有的列和列的 SQL 数据类型。这些 SQL 类型（VARCHAR、INTEGER 等）被映射到 Java 数据类型，在 MapReduce 应用中将使用这些对应的 Java 数据类型来保存字段的值。Sqoop 的代码生成器使用这些信息来创建对应表的类，用于保存从表中抽取的记录信息。

Sqoop 启动的 MapReduce 作业用到一个 InputFormat 类，通过 JDBC 从一个数据库中读取表的部分内容。Hadoop 提供的 DataDrivenDBInputFormat 能够为几个 Map 任务对查询结果进行划分。然而，人们经常将这样的查询划分到多个节点上执行，是为了更好地显示 Sqoop 的导入性能。查询时根据一个"划分列"（Splitting Column）来进行划分。Sqoop 根据表的元数据会选择一个合适的列作为划分列，通常情况下是表的主键。表的主键中的最小值和最大值会被读出，与目标任务数一起用来确定每个 Map 任务要执行的查询。

在生成反序列化代码和配置 InputFormat 之后，Sqoop 将作业发送到 MapReduce 集群。Map 任务执行查询，并且将 ResultSet 中的数据反序列化到生成类的实例，这些数据要么被直接保存在 SequenceFile 文件中，要么写到 HDFS 之前被转化成分隔的文本。

9.2.1 导入实例

本实例使用 Sqoop 把 MySQL 中的数据导入到 HDFS 上。在整个导入数据的过程中，要保证 MySQL 服务和 Hadoop 集群启动，且运行正常。用以下命令来查看这两个服务是否正常运行。

```
[zkpk@master ~]$ /etc/init.d/mysqld status
mysqld (pid  2340) is running...
[zkpk@master ~]$ jps
3559 ResourceManager
6221 Jps
3242 NameNode
3394 SecondaryNameNode
```

在使用 Sqoop 将数据导入到 HDFS 之前，先准备数据。以下操作都是在 MySQL 交互客户端执行，也是在为导入数据做的准备工作。

首先，通过命令登录 MySQL。

```
[zkpk@master ~]$ mysql -uhadoop -phadoop
```

然后，创建数据库。先查看数据库 test 是否存在。

```
mysql> show databases;
+--------------------+
| Database           |
+--------------------+
| information_schema |
| hive_13            |
| mysql              |
| test               |
+--------------------+
4 rows in set (0.02 sec)
```

上面查看的结果是 test 数据库已经存在了。如果不存在，则需要用如下语句来创建。

```
mysql> CREATE DATABASE test;
```

最后，需要在 MySQL 中创建表 sougou_example。

```
mysql> CREATE TABLE `test`.`sougou_example`(
    -> `uid` varchar(255) DEFAULT NULL,
    -> `cnt` int(11) DEFAULT NULL
    -> )
    -> ENGINE=InnoDB DEFAULT CHARSET=utf8;
Query OK, 0 rows affected (0.31 sec)
```

注意：语句中的引号是反引号`，而不是单引号'。

在 MySQL 中建表成功后，退出 MySQL。

表 sougou_example 中的数据如下：

```
11e2e89dbf484ed187e73cbeaf1e0084        1
4a6f0d5cc0bcf16e32e74ae49663b60d        1
4c4c23ff94387248f4dc88166177058a        1
596444b8c02b7b30c11273d5bbb88741        1
637b29b47fed3853e117aa7009a4b621        1
63fd6f826a5f83d795f08778468d0e14        1
698956eb07815439fe5f46e9a4503997        1
9c89762b968568aaa0bed63579088f8e        1
ec0363079f36254b12a5e30bdc070125        1
f31f594bd1f3147298bd952ba35de84d        1
```

在 MySQL 中的准备工作已经完成。把 MySQL 中的数据导入到 HDFS 上的命令如下：

```
[zkpk@master ~]$ cd ~/sqoop-1.4.5.bin_hadoop-2.0.4-alpha
[zkpk@master sqoop-1.4.5.bin_hadoop-2.0.4-alpha]$ bin/sqoop import --connect jdbc:mysql://master:3306/test --username hadoop --password hadoop --table sogou_example -m 1
```

Sqoop 的 import 工具会运行一个 MapReduce 作业，这个 MapReduce 作业会连接 MySQL 数据库，并读取表中的数据。

在默认情况下，该作业会并行使用 4 个 Map 任务来加速导入过程。每个任务都会将其所导入的数据写到一个单独的文件中，但所有 4 个文件都位于同一个目录中。在本例中，指定 Sqoop 只使用一个 Map 任务(-m 1)，这样只得到一个保存在 HDFS 中的文件。

9.2.2 导入数据的使用

一旦数据被导入到 HDFS 上，数据就可以供定制的 MapReduce 程序使用。导入的文本格式数据可以供 Hadoop Streaming 中的脚本或以 TextInputFormat 为默认格式运行的 MapReduce 作业使用。

Sqoop 生成的表类能够自动对字段分隔符(以及转义/包围字符)进行解析，抽出字段的值，并转换为相应的数据类型。这样，用户完全可以把精力集中在真正要运行的 MapReduce 作业上。而且，Sqoop 每个自动生成的类都有几个名为 parse() 的重载方法，这些方法可以对表示为 Text、CharSequence、char[] 或其他常见类型的数据进行操作。

提交上述 Map 任务在集群上运行时，Sqoop 通过 $HADOOP_CLASSPATH 确保文件所在的位置。运行之后，通过命令可以查看在 HDFS 的路径中有一个名为 part-m-00000 的文件，并且可以查看文件中的内容。命令如下：

```
[zkpk@master ~]$ hadoop fs -ls /user/zkpk/sougou_example
```

```
[zkpk@master ~]$ hadoop fs -cat /user/zkpk/sougou_example/part-m-00000
[zkpk@master sqoop-1.4.5.bin__hadoop-2.0.4-alpha]$ hadoop fs -ls /user/zkpk/soug
ou_example
-rw-r--r--   1 zkpk supergroup          0 2016-04-13 20:42 /user/zkpk/sougou_exa
mple/_SUCCESS
-rw-r--r--   1 zkpk supergroup        350 2016-04-13 20:42 /user/zkpk/sougou_exa
mple/part-m-00000
[zkpk@master sqoop-1.4.5.bin__hadoop-2.0.4-alpha]$ hadoop fs -cat /user/zkpk/sou
gou_example/part-m-00000
63fd6f826a5f83d795f08778468d0e14,1
596444b8c02b7b30c11273d5bbb88741,1
ec0363079f36254b12a5e30bdc070125,1
637b29b47fed3853e117aa7009a4b621,1
f31f594bd1f3147298bd952ba35de84d,1
11e2e89dbf484ed187e73cbeaf1e0084,1
4a6f0d5cc0bcf16e32e74ae49663b60d,1
4c4c23ff94387248f4dc88166177058a,1
698956eb07815439fe5f46e9a4503997,1
9c89762b968568aaa0bed63579088f8e,1
```

注意：在这个 MapReduce 示例的程序中，一个对象从 mapper 被发送到 reducer。这个自动生成的类实现了 Hadoop 提供的 Writable 接口，该接口允许通过 Hadoop 的序列化机制来发送对象，以及写到 SequenceFile 文件，或从 SequenceFile 文件读出对象。

9.2.3 数据导入代码生成

Sqoop 除了能把数据库 MySQL 表的内容写到 HDFS 上，同时还生成了一个 Java 源文件，保存在当前的本地目录中。在使用 Sqoop 导入数据时，可以看到 Sqoop 在将源数据库的表数据写到 HDFS 之前，会首先用生成的代码对其进行反序列化。

在导入过程中，由 Sqoop 生成的类会将每一条被导入的行保存在 SequenceFile 文件的键-值对格式中的"值"的位置。生成的类中能够保存一条从被导入表中抽取的记录，该类可以在 MapReduce 中使用这条记录，也可以将这条记录保存在 HDFS 中的一个 SequenceFile 文件中。

还可以使用另外一个 Sqoop 工具 codegen 来生成源代码，它不执行完整的导入操作，但生成的代码仍然会检查数据库表，以确定与每个字段相匹配的数据类型。如果用户意外地删除了生成的源代码，或希望使用不同于导入过程的设定来生成代码，都可以用这个工具来重新生成代码。

如果计划使用导入到 SequenceFile 文件中的记录，那么必须用到生成的类。因为对 SequenceFile 文件中的数据需要进行反序列化。如果使用的是文本文件中的记录，则不需要用到生成的代码。

下面是导入的代码，部分代码的解释是：

```
[zkpk@master ~]$cd ~/sqoop-1.4.5.bin_hadoop-2.0.4-alpha
[zkpk@master sqoop-1.4.5.bin_hadoop-2.0.4-alpha]$ bin/sqoop import --connect
jdbc:mysql://master:3306/test --username hadoop --password hadoop --table sougou_
example -m 1
```

其中，bin/sqoop import 表示从 MySQL 中把数据导入到 HDFS 上；--connect jdbc:mysql://master:3306/test 表示连接数据库 test 和 HDFS；--username hadoop 表示数据库 MySQL 的用户名；--password hadoop 表示数据库 MySQL 的用户密码；--table sougou_ example 表示需要导入到 HDFS 上的 MySQL 表；-m 1 表示 map 的个数。

注意：IP 部分需要使用 HadoopMaster 节点对应的 IP 地址或者节点名称。

9.3 数据导出

Sqoop 导出功能的架构与它的导入功能非常相似。在执行导出操作之前，Sqoop 会根据数据库连接字符串来选择一个导出方法。对于大多数系统来说，Sqoop 都会选择 JDBC。

Sqoop 的导出过程如图 9-2 所示。Sqoop 会根据目标表的定义生成一个 Java 类，这个生成的 Java 类能够从文本文件中解析出记录，并且能够向表中插入类型合适的值。接着会启动一个 MapReduce 作业，从 HDFS 中读取源数据文件，使用生成的 Java 类解析出记录信息，并且执行选定的导出方法。

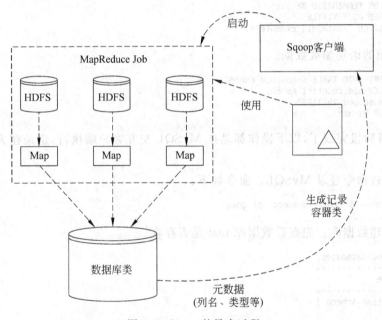

图 9-2 Sqoop 的导出过程

对于传统的数据库 MySQL 来说，Sqoop 可以使用 mysqlimport 的直接模式的方法。每个 Map 任务会生成一个 mysqlimport 进程，该进程通过本地文件系统上的一个命名 FIFO 通道流入 mysqlimport，然后再写入数据库。

虽然从 HDFS 上读取数据的 MapReduce 作业大多根据所处理文件的数量和大小来选择并行度（Map 任务的数量），但 Sqoop 的导出工具允许用户明确设定任务的数量。由于 Sqoop 导出性能会受并行的数据库写入线程数量的影响，因此 Sqoop 使用 CombineFileInputFormat 类将输入文件分组分配给少数几个 map 任务去执行。

9.3.1 导出实例

本小节中，使用 Sqoop 把 HDFS 中表 uid_example 的数据导出到 MySQL 中。

在整个导出数据的过程中，要保证 MySQL 服务和 Hadoop 集群启动且运行正常。用以下命令来查看这两个服务是否正常运行。

```
[zkpk@master ~]$ /etc/init.d/mysqld status
mysqld (pid  2340) is running...
[zkpk@master ~]$ jps
3559 ResourceManager
6221 Jps
3242 NameNode
3394 SecondaryNameNode
```

可以看出，MySQL 服务和 Hadoop 集群都运行正常。

在使用 Sqoop 将数据导出到 MySQL 之前，先准备数据。先在 Hive 中创建表 uid_example。

```
hive> CREATE TABLE sogou.uid_example(
    > uid STRING,
    > cnt INT)
    > ROW FORMAT DELIMITED
    > FIELDS TERMINATED BY '\t'
    > STORED AS TEXTFILE
    > LOCATION '/data/uid_example';
```

通过下面的语句加载数据。

```
hive> INSERT INTO TABLE sogou.uid_example
    > SELECT uid,count(*) AS cnt
    > FROM sougou_20111230
    > GROUP BY uid;
```

数据准备阶段完成了，以下操作都是在 MySQL 交互客户端执行，也是在为导数据做的准备工作。

首先，通过命令登录 MySQL。命令如下：

```
[zkpk@master ~]$ mysql -uhadoop -phadoop
```

然后，创建数据库。先查看数据库 test 是否存在。

```
mysql> show databases;
+--------------------+
| Database           |
+--------------------+
| information_schema |
| hive_13            |
| mysql              |
| test               |
+--------------------+
4 rows in set (0.02 sec)
```

上面查看的结果是 test 数据库已经存在了，如果不存在，则需要用如下语句来创建。

```
mysql> CREATE DATABASE test;
```

最后，在 MySQL 中创建表 sougou_example。

```
mysql> CREATE TABLE 'test'.'sougou_example'(
    -> 'uid' varchar(255) DEFAULT NULL,
    -> 'cnt' int(11) DEFAULT NULL
    -> )
    -> ENGINE=InnoDB DEFAULT CHARSET=utf8;
Query OK, 0 rows affected (0.31 sec)
```

把 HDFS 上的表导入到 MySQL 中。

```
[zkpk@master mapreduce]$ cd /home/zkpk/sqoop-1.4.5.bin__hadoop-2.0.4-alpha
```

```
[zkpk@master sqoop-1.4.5.bin__hadoop-2.0.4-alpha]$ bin/sqoop export --connect
jdbc:mysql//192.168.112.128:3306/test --username hadoop --password hadoop --
table sougou_example --export-dir '/data/uid_example' --fields-terminated-by '\t'
```

其中,bin/sqoop export 表示将数据从 HDFS 上导出到 MySQL 中;--export-dir '/data/uid_example'表示 Hive 中被导出的文件(Hive 表存放在 HDFS 上);--fields-terminated-by '\t':Hive 中被导出的文件字段的分隔符。

导入成功后,进入 MySQL 查看。

```
mysql> select * from sougou_example;
+----------------------------------+------+
| uid                              | cnt  |
+----------------------------------+------+
| 63fd6f826a5f83d795f08778468d0e14 |    1 |
| 596444b8c02b7b30c11273d5bbb88741 |    1 |
| ec0363079f36254b12a5e30bdc070125 |    1 |
| 637b29b47fed3853e117aa7009a4b621 |    1 |
| f31f594bd1f3147298bd952ba35de84d |    1 |
| 11e2e89dbf484ed187e73cbeaf1e0084 |    1 |
| 4a6f0d5cc0bcf16e32e74ae49663b60d |    1 |
| 4c4c23ff94387248f4dc88166177058a |    1 |
| 698956eb07815439fe5f46e9a4503997 |    1 |
| 9c89762b968568aaa0bed63579088f8e |    1 |
+----------------------------------+------+
10 rows in set (0.03 sec)
```

导出成功后的数据在 MySQL 中,用户可以通过自己的需求来操作此表,如表的查询等。

9.3.2 导出和SequenceFile

9.3.1 小节的导出示例是从一个 Hive 表中读取源数据,该 Hive 表以分隔文本文件形式保存在 HDFS 中。Sqoop 也可以从非 Hive 表的分隔文本文件中导出数据。Sqoop 不仅可以导出 MapReduce 作业结果的文本文件,还可以将存储在 SequenceFile 中的记录导出到输出表,但是有一些限制。SequenceFile 中可以保存任意类型的记录,但由于 Sqoop 的导出工具从 SequenceFile 中读取对象后直接发送到 OutputCollector,再由它将这些对象传递给数据库导出 OutputFormat。所以记录必须被保存在 SequenceFile 键-值对的"值"部分,并且必须继承抽象类 com.cloudera.sqoop.lib.SqoopRecord,就像 Sqoop 生成的所有类那样。

如果基于导出目标表使用 codegen 工具(sqoop-codegen)为记录生成一个 SqoopRecord 的实现,那就可以写一个 MapReduce 应用程序,填充这个类的实例,并将它们写入 SequenceFile,使用 sqoop-export 将这些 SequenceFile 文件导出到表中。还可以将数据放入 SqoopRecord 实例中,然后保存在 SequenceFile 文件中。如果数据是从关系型数据库表导入 HDFS 的,那么在经过某种形式的修改后,可以将结果保存在持有相同数据类型记录的 SequenceFile 文件中。

本章小结

本章首先学习了 Sqoop 的安装部署,然后介绍了 Sqoop 的导入导出过程的工作原理,并用实例说明如何使用 Sqoop 把 MySQL 中的数据导入到 HDFS 上,把 HDFS 上的数据导出到 MySQL。

习 题

1. 选择题

(1) 在安装部署 Sqoop 时,下列不需要在配置环境变量中配置的一项是(　　)。
  A. JAVA_HOME   B. HADOOP_COMMON_HOME
  C. HADOOP_MAPRED_HOME   D. HIVE_HOME

(2) 把 MySQL 中的数据导入到 HDFS 的过程中,下列叙述不正确的是(　　)。
  A. Sqoop 启动的 MapReduce 作业用到一个 InputFormat 类,通过 JDBC 从一个数据库中读取表的部分内容
  B. Hadoop 提供的 DataDrivenDBInputFormat 能够为几个 Map 任务对查询结果进行划分
  C. Sqoop 根据表的元数据会选择一个合适的列作为划分列,通常情况下是表的外键
  D. 在生成反序列化代码和配置 InputFormat 之后,Sqoop 将作业发送到 MapReduce 集群

(3) 下列选项中,(　　)在 Sqoop 导入导出时用不到。
  A. --connect   B. --table
  C. --export-dir   D. --import

(4) 关于使用 Sqoop 把 HDFS 上的数据导出到 MySQL 的过程中,下列说法中正确的是(　　)。
  A. 在使用 Sqoop 将数据导出到 MySQL 之前,要在 Hive 中创建表,表中不需要数据
  B. 要保证 MySQL 服务和 Hadoop 集群启动且运行正常
  C. 在 MySQL 数据库中创建表时,表的名字和字段需要用引号引起来
  D. 以上叙述都正确

2. 问答题

(1) 简述 Sqoop 组件在生态圈中担任的角色。
(2) 画出 Sqoop 的导入过程并简单描述它的工作原理。
(3) 画出 Sqoop 的导出过程并简单描述它的工作原理。

第 10 章

Hadoop 工作流引擎 Oozie

本章摘要

在 Hadoop 中执行的任务有时需要把多个 MapReduce 作业连接到一起,这样才能达到目的。在 Hadoop 生态圈中,Oozie 组件使用户可以把多个 MapReduce 作业组合到一个逻辑工作单元中,从而完成更大型的任务。

本章将首先介绍 Oozie 相关的基本概念,并给出 Oozie 安装的详细步骤,然后将介绍 Oozie 的使用,实现作业的部署与执行,最后介绍 Oozie 的控制台及一些高级特性。

10.1 Oozie 是什么

1. 工作流是什么

工作流(Work Flow)就是工作流程的计算模型,即将工作流程中的工作如何前后组织在一起的逻辑和规则,在计算机中以恰当的模型进行表示,并对其实施计算。

2. Oozie 简介

Oozie 是 Apache 项目,由雅虎开发出来。它是一个 Hadoop 工作流引擎,用于管理数据加工活动。Oozie 可以用于运行 Hadoop MapReduce 和 Pig 任务工作流,同时 Oozie 还是一个 Java Web 程序,运行在 Java Servlet 容器中,如 Tomcat,并使用数据库来存储工作流和当前运行的工作流实例(实例的状态和变量)。

Oozie 工作流中拥有多个 Action,如 Hadoop MapReduce Job、Hadoop Pig Job 等,所有的 Action 以有向无环图(Direct Acyclic Graph,DAG)的模式部署运行。因此,Action 的运行步骤是有方向的,只能等上一个 Action 运行完成后,才能运行下一个 Action。

10.2 Oozie 的安装

1. 安装 maven 环境

(1) 确定安装了 JDK(JDK 版本不能超过 1.7,这里以 1.7 版本为例)。

```
[zkpk@master ~]$ java -version
java version "1.7.0_71"
Java(TM) SE Runtime Environment (build 1.7.0_71-b14)
Java HotSpot(TM) 64-Bit Server VM (build 24.71-b01, mixed mode)
```

(2) 在 root 用户下安装 yum -y install apache-maven,可能会出现如下错误。

```
[root@master ~]# yum -y install apache-maven
Loaded plugins: fastestmirror, refresh-packagekit, security
Loading mirror speeds from cached hostfile
 * base: mirrors.tuna.tsinghua.edu.cn
 * extras: mirrors.tuna.tsinghua.edu.cn
 * updates: mirrors.tuna.tsinghua.edu.cn
Setting up Install Process
No package apache-maven available.
Error: Nothing to do
```

(3) 这时需要用 wget 命令去网络上下载资源包。

```
[root@master ~]#wget http://repos.fedorapeople.org/repos/dchen/apache-maven/epel-apache-maven.repo -O /etc/yum.repos.d/epel-apache-maven.repo
```

然后重新安装 maven。

```
[root@master ~]#yum -y install apache-maven
Dependency Installed:
  giflib.x86_64 0:4.1.6-3.1.el6
  java-1.6.0-openjdk.x86_64 1:1.6.0.40-1.13.12.6.el6_8
  java-1.7.0-openjdk.x86_64 1:1.7.0.111-2.6.7.2.el6_8
  java-1.7.0-openjdk-devel.x86_64 1:1.7.0.111-2.6.7.2.el6_8
  jpackage-utils.noarch 0:1.7.5-3.16.el6
  lksctp-tools.x86_64 0:1.0.10-7.el6
  pcsc-lite-libs.x86_64 0:1.5.2-15.el6
  ttmkfdir.x86_64 0:3.0.9-32.1.el6
  tzdata-java.noarch 0:2016h-1.el6
  xorg-x11-fonts-Type1.noarch 0:7.2-11.el6
```

(4) 输入 mvn -version,查看 maven 的安装路径时,JDK 的版本也同样显示出来。

```
[root@master zkpk]# mvn -version
Apache Maven 3.3.9 (bb52d8502b132ec0a5a3f4c09453c07478323dc5; 2015-11-10T08:41:47-08:00)
Maven home: /usr/share/apache-maven
Java version: 1.7.0_71, vendor: Oracle Corporation
Java home: /usr/java/jdk1.7.0_71/jre
Default locale: en_US, platform encoding: UTF-8
OS name: "linux", version: "2.6.32-431.el6.x86_64", arch: "amd64", family: "unix"
```

如果以上 4 步顺利通过,则说明 maven 环境安装好了。接下来可以正式进行 Oozie 的安装。

2. 安装 Oozie

(1) 解压 Oozie 4.2.0(普通用户下)。

```
[zkpk@master ~]$tar -zxvf oozie-4.2.0.tar.gz
[zkpk@master ~]$cd oozie-4.2.0
```

(2) 编译,进入 Oozie 解压缩目录,使用以下命令。

```
[zkpk@master oozie-4.2.0]$bin/mkdistro.sh -DskipTests
```

可能会出现如下信息,是因为 http://repository.codehaus.org/ 连接失败。

```
[ERROR] Failed to execute goal org.apache.maven.plugins:maven-site-plugin:2.0-be
ta-6:site (default) on project oozie-docs: The site descriptor cannot be resolve
d from the repository: Could not transfer artifact org.apache:apache:xml:site_en
:16 from/to Codehaus repository (http://repository.codehaus.org/): repository.co
dehaus.org: Name or service not known
[ERROR] org.apache:apache:xml:16
[ERROR]
[ERROR] from the specified remote repositories:
[ERROR] central (http://repo1.maven.org/maven2, releases=true, snapshots=false),
[ERROR] Codehaus repository (http://repository.codehaus.org/, releases=true, sna
pshots=false)
```

遇到上述错误信息,需要修改配置文件 pom.xml。

```
[zkpk@master oozie-4.2.0]$gedit pom.xml
```

找到如下两行配置,并进行修改。

```
<id>Codenhaus repository</id>
<url>http://repository.codehaus.org/</url>
```

把上面内容修改为:

```
<id>Codehaus repository</id>
<url>https://repository-master.mulesoft.org/nexus/content/groups/public/</url>
```

如果遇到如图 10-1 所示信息,表示运行 Maven 的 JVM 已经耗尽内存。

```
[INFO] Apache Oozie WebApp ............................... FAILURE [ 4.126 s]
[INFO] Apache Oozie Tools ................................ SKIPPED
[INFO] Apache Oozie MiniOozie ............................ SKIPPED
[INFO] Apache Oozie Distro ............................... SKIPPED
[INFO] Apache Oozie ZooKeeper Security Tests ............. SKIPPED
[INFO] ------------------------------------------------------------------------
[INFO] BUILD FAILURE
[INFO] ------------------------------------------------------------------------
[INFO] Total time: 07:55 min
[INFO] Finished at: 2016-12-25T05:03:12-08:00
[INFO] Final Memory: 62M/452M
[INFO] ------------------------------------------------------------------------
[ERROR] PermGen space -> [Help 1]
[ERROR]
[ERROR] To see the full stack trace of the errors, re-run Maven with the -e swit
ch.
[ERROR] Re-run Maven using the -X switch to enable full debug logging.
[ERROR]
[ERROR] For more information about the errors and possible solutions, please rea
d the following articles:
[ERROR] [Help 1] http://cwiki.apache.org/confluence/display/MAVEN/OutOfMemoryErr
or

ERROR, Oozie distro creation failed
```

图 10-1 运行 Maven 的 JVM 已经耗尽内存

需要修改配置文件信息,使用 root 用户编辑以下文件。

```
[root@master ~]#vim /etc/profile
```

添加下面一行信息,然后保存退出。

```
export MAVEN_OPTS="-Xmx512m -XX:MaxPermSize=128m"
```

然后重新编译。

```
[zkpk@master oozie-4.2.0]$bin/mkdistro.sh -DskipTests
```

如果出现如图 10-2 所示结果,则表示编译成功。

```
[INFO] Reactor Summary:
[INFO]
[INFO] Apache Oozie Main ................................. SUCCESS [  2.431 s]
[INFO] Apache Oozie Hadoop Utils ......................... SUCCESS [  1.371 s]
[INFO] Apache Oozie Hadoop Distcp hadoop-1-4.2.0 ......... SUCCESS [  0.095 s]
[INFO] Apache Oozie Hadoop Auth hadoop-1-4.2.0 ........... SUCCESS [  0.235 s]
[INFO] Apache Oozie Hadoop Libs .......................... SUCCESS [  0.021 s]
[INFO] Apache Oozie Client ............................... SUCCESS [01:43 min]
[INFO] Apache Oozie Share Lib Oozie ...................... SUCCESS [  2.548 s]
[INFO] Apache Oozie Share Lib HCatalog ................... SUCCESS [  3.230 s]
[INFO] Apache Oozie Share Lib Distcp ..................... SUCCESS [  0.619 s]
[INFO] Apache Oozie Core ................................. SUCCESS [01:21 min]
[INFO] Apache Oozie Share Lib Streaming .................. SUCCESS [  7.992 s]
[INFO] Apache Oozie Share Lib Pig ........................ SUCCESS [  5.820 s]
[INFO] Apache Oozie Share Lib Hive ....................... SUCCESS [  5.072 s]
[INFO] Apache Oozie Share Lib Hive 2 ..................... SUCCESS [  4.552 s]
[INFO] Apache Oozie Share Lib Sqoop ...................... SUCCESS [  3.151 s]
[INFO] Apache Oozie Examples ............................. SUCCESS [  6.571 s]
[INFO] Apache Oozie Share Lib Spark ...................... SUCCESS [  8.376 s]
[INFO] Apache Oozie Share Lib ............................ SUCCESS [ 17.596 s]
[INFO] Apache Oozie Docs ................................. SUCCESS [ 45.103 s]
[INFO] Apache Oozie WebApp ............................... SUCCESS [03:41 min]
[INFO] Apache Oozie Tools ................................ SUCCESS [  5.992 s]
[INFO] Apache Oozie MiniOozie ............................ SUCCESS [  1.793 s]
[INFO] Apache Oozie Distro ............................... SUCCESS [06:59 min]
[INFO] Apache Oozie ZooKeeper Security Tests ............. SUCCESS [ 10.156 s]
[INFO] ------------------------------------------------------------------------
[INFO] BUILD SUCCESS
[INFO] ------------------------------------------------------------------------
```

图 10-2 编译成功

编译后的文件在 distro/target 文件夹,这里的文件名为 oozie-4.2.0-distro.tar.gz,如下所示。

```
[zkpk@master target]$ ls
antrun         maven-archiver                   oozie-4.2.0-distro.tar.gz   tomcat
archive-tmp    maven-shared-archive-resources   oozie-distro-4.2.0.jar
classes        oozie-4.2.0-distro               test-classes
```

(3) 安装 Oozie server。

经过上面的编译,得到了二进制版的 Oozie,下面就可以部署 server 了。由此可见,Oozie 使用的是 B/S 模式。解压 oozie-4.2.0-distro.tar.gz 这个编译后的文件,然后进入解压后的文件目录,并创建文件夹 libext。

```
[zkpk@master target]$tar -zxvf oozie-4.2.0-distro.tar.gz
[zkpk@master target]$cd oozie-4.2.0-distro/oozie-4.2.0
[zkpk@master oozie-4.2.0]$mkdir libext
```

把 ext-2.2.zip 复制到 libext 目录下,其中 ext-2.2.zip 是 Oozie server 需要的一个 js 库,可以去官网 http://docs.sencha.com/ 查找。然后,把 Hadoop 的一些 jar 包也放到 libext 文件夹内。

```
[zkpk@master oozie-4.2.0]$cp ${HADOOP_HOME}/share/hadoop/*/*.jar libext/
[zkpk@master oozie-4.2.0]$cp ${HADOOP_HOME}/share/hadoop/*/lib/*.jar libext/
```

把 Hadoop 与 Tomcat 冲突的 jar 包去掉,如下所示。

```
[zkpk@master libext]$rm jasper-compiler-5.5.23.jar
```

```
[zkpk@master libext]$rm jasper-runtime-5.5.23.jar
[zkpk@master libext]$rm jsp-api-2.1.jar
```

Oozie server 还需要依赖数据库,这里用的是 MySQL,需要把 MySQL 的驱动程序 jar 包也放在 libext 中。切换到 mysql-connector-java-5.1.27 目录下,然后将 jar 包复制到 libext 中。

```
[zkpk@master mysql-connector-java-5.1.27]$cp mysql-connector-java-5.1.27-bin.jar ~/oozie-4.2.0/distro/target/oozie-4.2.0-distro/oozie-4.2.0/libext/
```

修改 conf/oozie-site.xml 的内容如下:

```
[zkpk@master ~]$cd oozie-4.2.0/distro/target/oozie-4.2.0-distro/oozie-4.2.0/conf
[zkpk@master conf]$vim oozie-site.xml

<property>
    <name>oozie.service.JPAService.create.db.schema</name>
    <value>true</value>
</property>
<property>
    <name>oozie.service.JPAService.jdbc.driver</name>
    <value>com.mysql.jdbc.Driver</value>
</property>
<property>
    <name>oozie.service.JPAService.jdbc.url</name>
    <value>jdbc:mysql://master:3306/oozie?createDatabaseIfNotExist=true</value>
</property>
<property>
    <name>oozie.service.JPAService.jdbc.username</name>
    <value>hadoop</value>
</property>
<property>
    <name>oozie.service.JPAService.jdbc.password</name>
    <value>hadoop</value>
</property>
<property>
    <name>oozie.service.HadoopAccessorService.hadoop.configurations</name>
    <value>*=/home/zkpk/hadoop-2.7.2/etc/hadoop</value>
</property>
```

上面的配置中,指定了 Hadoop 配置文件位置:*=/home/zkpk/hadoop-2.7.2/etc/hadoop。

注意:这里的 *= 不能缺少,根据实际情况修改 Hadoop 路径即可。

(4) 打 war 包。

```
[zkpk@master ~]$cd oozie-4.2.0/distro/target/oozie-4.2.0-distro/oozie-4.2.0/bin
[zkpk@master bin]$./oozie-setup.sh prepare-war
```

截取最后的执行结果,如下所示。

```
INFO: Adding extension: /home/zkpk/oozie-4.2.0/distro/target/oozie-4.2.0-distro/
oozie-4.2.0/libext/slf4j-api-1.7.5.jar
INFO: Adding extension: /home/zkpk/oozie-4.2.0/distro/target/oozie-4.2.0-distro/
oozie-4.2.0/libext/slf4j-log4j12-1.7.5.jar
INFO: Adding extension: /home/zkpk/oozie-4.2.0/distro/target/oozie-4.2.0-distro/
oozie-4.2.0/libext/snappy-java-1.0.4.1.jar
INFO: Adding extension: /home/zkpk/oozie-4.2.0/distro/target/oozie-4.2.0-distro/
oozie-4.2.0/libext/stax-api-1.0-2.jar
INFO: Adding extension: /home/zkpk/oozie-4.2.0/distro/target/oozie-4.2.0-distro/
oozie-4.2.0/libext/xmlenc-0.52.jar
INFO: Adding extension: /home/zkpk/oozie-4.2.0/distro/target/oozie-4.2.0-distro/
oozie-4.2.0/libext/xz-1.0.jar
INFO: Adding extension: /home/zkpk/oozie-4.2.0/distro/target/oozie-4.2.0-distro/
oozie-4.2.0/libext/zookeeper-3.4.6.jar

New Oozie WAR file with added 'ExtJS library, JARs' at /home/zkpk/oozie-4.2.0/di
stro/target/oozie-4.2.0-distro/oozie-4.2.0/oozie-server/webapps/oozie.war

INFO: Oozie is ready to be started
```

若如上所示,则说明 server 已经生成了。

(5) 修改 HDFS 配置。

```
[zkpk@master bin]$ vim ~/hadoop-2.7.2/etc/hadoop/core-site.xml
```

修改 Hadoop 中的配置文件 core-site.xml,内容如下:

```
<property>
    <name>hadoop.proxyuser.zkpk.hosts</name>
    <value>master</value>
</property>
<property>
    <name>hadoop.proxyuser.zkpk.groups</name>
    <value>zkpk</value>
</property>
```

其中,name 中的 zkpk 是用户名,value 中的 master 是主机名,zkpk 是用户组名。

(6) 上传 jar 包到 HDFS 上(前提是 Hadoop 已正常启动,此处 192.168.111.128 为本机 IP,请自行替换成自己的 IP)。

```
[zkpk@master oozie-4.2.0]$ bin/oozie-setup.sh sharelib create -fs hdfs://192.168
.111.128:9000
  setting CATALINA_OPTS="$CATALINA_OPTS -Xmx1024m"
SLF4J: Class path contains multiple SLF4J bindings.
SLF4J: Found binding in [jar:file:/home/zkpk/oozie-4.2.0/distro/target/oozie-4.2
.0-distro/oozie-4.2.0/lib/slf4j-log4j12-1.6.6.jar!/org/slf4j/impl/StaticLoggerBi
nder.class]
SLF4J: Found binding in [jar:file:/home/zkpk/oozie-4.2.0/distro/target/oozie-4.2
.0-distro/oozie-4.2.0/lib/slf4j-simple-1.6.6.jar!/org/slf4j/impl/StaticLoggerBin
der.class]
SLF4J: Found binding in [jar:file:/home/zkpk/oozie-4.2.0/distro/target/oozie-4.2
.0-distro/oozie-4.2.0/libext/slf4j-log4j12-1.7.5.jar!/org/slf4j/impl/StaticLogge
rBinder.class]
SLF4J: See http://www.slf4j.org/codes.html#multiple_bindings for an explanation.
SLF4J: Actual binding is of type [org.slf4j.impl.Log4jLoggerFactory]
the destination path for sharelib is: /user/zkpk/share/lib/lib_20161031065144
```

（7）启动 Oozie（MySQL 已正常启动）。

```
[zkpk@master oozie-4.2.0]$bin/oozied.sh start
```

启动信息，如下所示。

```
Oozie DB has been created for Oozie version '4.2.0'

The SQL commands have been written to: /tmp/ooziedb-7953795009476274837.sql

Using CATALINA_BASE:   /home/zkpk/oozie-4.2.0/distro/target/oozie-4.2.0-distro/o
ozie-4.2.0/oozie-server
Using CATALINA_HOME:   /home/zkpk/oozie-4.2.0/distro/target/oozie-4.2.0-distro/o
ozie-4.2.0/oozie-server
Using CATALINA_TMPDIR: /home/zkpk/oozie-4.2.0/distro/target/oozie-4.2.0-distro/o
ozie-4.2.0/oozie-server/temp
Using JRE_HOME:        /usr/java/jdk1.7.0_71/
Using CLASSPATH:       /home/zkpk/oozie-4.2.0/distro/target/oozie-4.2.0-distro/o
ozie-4.2.0/oozie-server/bin/bootstrap.jar
Using CATALINA_PID:    /home/zkpk/oozie-4.2.0/distro/target/oozie-4.2.0-distro/o
ozie-4.2.0/oozie-server/temp/oozie.pid
```

进入默认端口 11000 后，进入如图 10-3 所示界面。

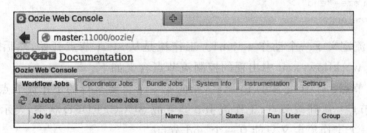

图 10-3　Oozie 安装完成界面

若如上所示，则 Oozie 安装完成。

10.3　Oozie 的编写与运行

Oozie 任务流包括 workflow、coordinator、bundle。workflow 描述任务执行顺序的 DAG（有向无环图）；coordinator 用于定时任务触发，相当于 workflow 的定时管理器；bundle 是对多个 coordinator 的抽象。

10.3.1　Workflow 组件

Oozie 定义了一种基于 XML 的 HPDL（Hadoop Process Definition Language）来描述 workflow 的 DAG。在 workflow 中定义了控制流节点（Control Flow Nodes）和动作节点（Action Nodes）。

其中，控制流节点定义了流程的开始（start）和结束（end），以及控制流程的执行路径（Execution Path），如 decision、fork、join 等；而动作节点包括 Hadoop 任务、SSH、HTTP、eMail 和 Oozie 子流程等。下面将以 Oozie 自带的 examples 为例。

1. 准备工作

`[zkpk@master ~]$ hadoop dfs -mkdir examples`

2. 工作流的介绍

`[zkpk@master ~]$ cd oozie-4.2.0/examples/target/oozie-examples-4.2.0-examples/examples/apps/cron`
`[zkpk@master map-reduce]$ vim workflow.xml`

将内容修改如下：

```
<workflow-app xmlns="uri:oozie:workflow:0.5" name="one-op-wf">
    <start to="action1"/>
    <action name="action1">
        <fs/>
    <ok to="end"/>
    <error to="end"/>
    </action>
    <end name="end"/>
</workflow-app>
```

这个工作流定义了一个动作 action1。工作流是从 start 节点开始，然后把控制权交给 action1 动作。workflow.xml 中出现的控制流节点含义如下。

（1）start 节点：start 元素的 to 属性，指向第一个将要执行的工作流节点。

（2）end 节点：达到该节点，工作流 Job 会变成 SUCCEEDED 状态，表示成功完成。需要注意的是，一个工作流定义只能有一个 end 节点。

（3）kill 节点：kill 元素的 name 属性是要杀死的工作流节点的名称，message 属性指定了工作流节点被杀死的备注信息。达到该节点，工作流 Job 会变成状态 KILLED。

（4）decision 节点：decision 节点通过预定义一组条件，当工作流 Job 执行到该节点时，会根据其中的条件进行判断选择，满足条件的路径将被执行。decision 节点通过 switch…case 语法来进行路径选择，只要有满足条件的判断，就会执行对应的路径，如果没有配置，则指向元素默认的节点。

（5）fork 节点和 join 节点：fork 节点下会有多个 path 元素，指定了可以并发执行的多个执行路径。fork 中多个并发执行路径会在 join 节点的位置会合，只有所有的路径都到达后，才会继续执行 join 节点。

动作节点是在工作流程定义中能够触发一个计算任务或者处理任务执行的节点，Oozie 内置支持的动作节点类型如下。

（1）MapReduce 动作：MapReduce 动作会在工作流 Job 中启动一个 MapReduce Job 任务运行，可以详细配置这个 MapReduce Job。另外，可以通过 MapReduce 元素的子元素来配置一些其他的任务，如 streaming、pipes、file、archive 等。

（2）Hive 动作：Hive 主要是基于类似 SQL 的 HQL 语言，它能够方便地操作 HDFS 中的数据，实现对海量数据的分析工作。

（3）Sqoop 动作：Sqoop 是一个能够在 Hadoop 和结构化存储系统之间进行数据导入导出的工具。

（4）Pig 动作：Pig 动作可以启动运行 Pig 脚本实现的 Job。

（5）Fs 动作：Fs 动作主要是基于 HDFS 的一些基本操作，如删除路径、创建路径、移动文件、设置文件权限等。

（6）SSH 动作：该动作主要是通过 SSH 登录到一台主机，能够执行一组 shell 命令。它在 Oozie schema 0.2 中已经被删除。

（7）Java 动作：Java 动作是执行一个具有 Main 入口方法的应用程序，在 Oozie 工作流定义中，会作为一个 MapReduce Job 执行，这个 Job 只有一个 Map 任务。用户需要指定 NameNode、JobTracker 的信息，配置一个 Java 应用程序的 JVM 选项参数(java-opts)，以及传给主函数(arg)。

（8）Sub-workflow 动作：Sub-workflow 动作是一个子流程的动作。在主流程执行过程中，遇到子流程节点执行时，会一直等待子流程节点执行完成后，才能继续跳转到下一个要执行的节点。

（9）Shell 动作：Shell 动作可以执行 Shell 命令，并通过配置来命令所需要的参数。

10.3.2　Coordinator 组件

如果现在有一个工作流 Job，希望每天 00:00 启动运行，那么可以通过写一个定时脚本来调度程序运行。Coordinator 能够将每个工作流 Job 作为一个动作(Action)来运行，相当于工作流定义中的一个执行节点(可以理解为工作流的工作流)，这样就能够将多个工作流 Job 组织起来，称为 Coordinator Job，并指定触发时间和频率，还可以配置数据集、并发数等。一个 Coordinator Job 包含了在 Job 外部设置执行周期和频率的语义，类似于在工作流外部增加了一个协调器来管理这些工作流 Job 的运行。

Coordinator 可用于指定触发 Workflow 作业执行的条件。这些条件可以作为数据可用性、时间或发起作业必须满足的外部事件。通过定义多个顺序运行的、前一个输出作为下一个输入的 Workflow，Coordinator 也可以定义常规运行的 Workflow 作业之间的依赖。

```
[zkpk@master ~]$ cd oozie-4.2.0/examples/target/oozie-examples-4.2.0-examples/
examples/apps/cron
[zkpk@master cron]$ vim coordinator.xml
```

coordinator.xml 配置文件如下：

```xml
<coordinator-app name="cron-coord" frequency="${coord:minutes(10)}" start="
${start}" end="${end}" timezone="UTC" xmlns="uri:oozie:coordinator:0.2">
    <action>
        <workflow>
            <app-path>${workflowAppUri}</app-path>
            <configuration>
                <property>
                    <name>jobTracker</name>
                    <value>${jobTracker}</value>
                </property>
                <property>
                    <name>nameNode</name>
                    <value>${nameNode}</value>
```

```xml
                </property>
                <property>
                    <name>queueName</name>
                    <value>${queueName}</value>
                </property>
            </configuration>
        </workflow>
    </action>
</coordinator-app>
```

在上面的配置文件中，coord：minutes（10）定义定时时间间隔是 10min。XML Coordinator 语言的主要元素如下。

(1) coordinator-app：这是 Coordinator 应用程序的顶层元素。

(2) controls：该元素指定 Coordinator 作业的执行策略。

(3) action：指定一个 Workflow 应用程序的位置和配置属性。

(4) dataset：表示一个逻辑名称所指定的一组数据。

(5) input-events：指定了提交 Coordinator 动作时需要的输入条件。

(6) ouput-events：指定了 Coordinator 动作应产生的数据集。

10.3.3 Bundle 组件

Bundle 的作用是将多个 Coordinator 管理起来，这样只需要提供一个 Bundle 提交即可，然后可以 start/stop/suspend/resume 任何 Coordinator。

Bundle 是顶层抽象，它允许将一组 Coordinator 应用程序打包成一个 Bundle 应用程序，将组成一个 Bundle 的多个 Coordinator 应用程序作为一个整体来进行控制。Bundle 在它的 Coordinator 应用程序之间不指定任何显示的依赖关系。这些依赖关系可以通过输入输出事件在 Coordinator 应用程序自身中指定。

XML bundle 语言的主要元素如下。

(1) bundle-app：Bundle 应用程序的顶层元素。

(2) name：用于指定 Bundle 的名字。

(3) controls：只包含一个属性 kick-off-time。用于指定 Bundle 应用程序的开始时间。

(4) coordinator：描述 Bundle 中包含的 Coordinator 应用程序。一个 Bundle 应用程序可以有多个 Coordinator 元素。

10.3.4 作业的部署与执行

本小节通过 Oozie 自带的例子来学习作业的部署与执行，需要修改配置文件，将文件上传到 HDFS 上，把作业提交到集群来进行处理，具体步骤如下。

(1) 修改 job.properties 文件。

```
[zkpk@master ~]$cd
oozie-4.2.0/examples/target/oozie-examples-4.2.0-examples/examples/apps/cron
[zkpk@master cron]$vim job.properties
```

添加如下内容。

```
nameNode=hdfs://localhost:9000
jobTracker=localhost:9001
queueName=default
examplesRoot=examples
```

(2) 把文件上传到 HDFS 上。

```
[zkpk@master ~]$cd
oozie-4.2.0/examples/target/oozie-examples-4.2.0-examples/
[zkpk@master oozie-examples-4.2.0-examples]$hadoop dfs -put examples/ examples
```

(3) 运行作业。

```
[zkpk@master ~]$cd
oozie-4.2.0/distro/target/oozie-4.2.0-distro/oozie-4.2.0
[zkpk@master oozie-4.2.0]$bin/oozie job -oozie
http://master:11000/oozie -config
/home/zkpk/oozie-4.2.0/examples/target/oozie-examples-4.2.0-examples/examples/
apps/cron/job.properties  -run
```

运行结果如下:

```
[zkpk@master oozie-4.2.0]$ bin/oozie job -oozie http://master:11000/oozie -confi
g /home/zkpk/oozie-4.2.0/examples/target/oozie-examples-4.2.0-examples/examples/
apps/cron/job.properties  -run
SLF4J: Class path contains multiple SLF4J bindings.
SLF4J: Found binding in [jar:file:/home/zkpk/oozie-4.2.0/distro/target/oozie-4.2
.0-distro/oozie-4.2.0/lib/slf4j-log4j12-1.6.6.jar!/org/slf4j/impl/StaticLoggerBi
nder.class]
SLF4J: Found binding in [jar:file:/home/zkpk/oozie-4.2.0/distro/target/oozie-4.2
.0-distro/oozie-4.2.0/lib/slf4j-simple-1.6.6.jar!/org/slf4j/impl/StaticLoggerBin
der.class]
SLF4J: Found binding in [jar:file:/home/zkpk/oozie-4.2.0/distro/target/oozie-4.2
.0-distro/oozie-4.2.0/libext/slf4j-log4j12-1.7.5.jar!/org/slf4j/impl/StaticLogge
rBinder.class]
SLF4J: See http://www.slf4j.org/codes.html#multiple_bindings for an explanation.
SLF4J: Actual binding is of type [org.slf4j.impl.Log4jLoggerFactory]
job: 0000000-161105195215852-oozie-zkpk-C
```

进入默认端口 11000 后,看到如图 10-4 所示界面(只截取了一部分)。

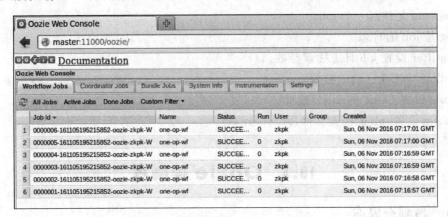

图 10-4　完成作业的部署与执行界面

以上说明完成作业的部署与执行。

10.3.5 向作业传递参数

[zkpk@master oozie-4.2.0]$bin/oozie help

向作业传递参数界面如图 10-5 所示。

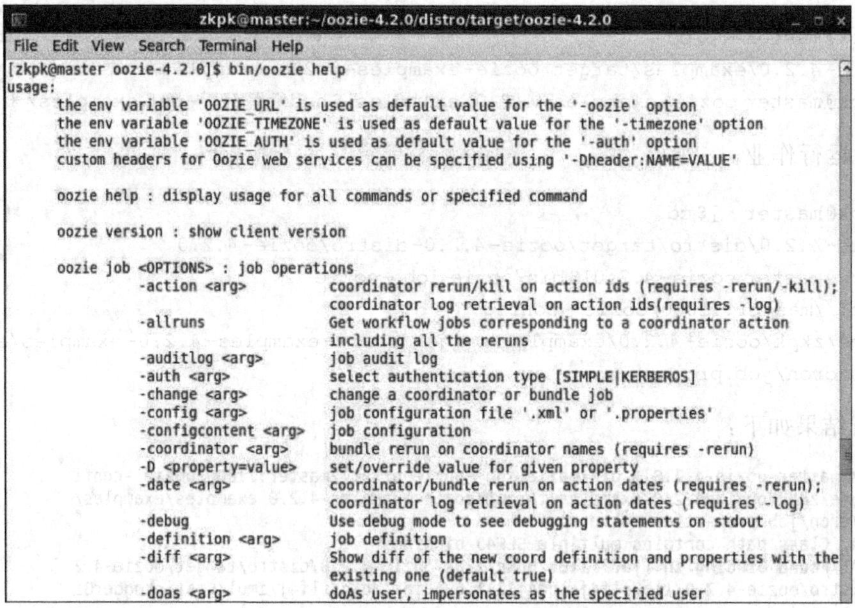

图 10-5 向作业传递参数界面

上面命令可以查看 job 执行参数、job 状态参数、管理员执行参数、Pig、Hive、Sqoop、MapReduce 等使用的参数设置，对部分关于 job 的命令解释如下。

-run 表示直接运行一个 job。

-submit 是提交 job。

-return 重新返回一个 job。

-suspend 是挂起一个 job。

-log 是 job 的日志。

-logfilter 设置 job 日志搜索参数。

-D 设置或覆盖已给的参数。

-oozie 设置 Oozie 的 URL 地址。

-config 设置 job 的配置文件(.properties)。

10.4 Oozie 控制台

10.4.1 控制台界面

以 10.3.4 小节的例子的结果为例，进入 http//master:11000/oozie/后显示如图 10-6 所示 Oozie 的控制台界面。Oozie 控制台显示首页，并默认选中 Workflow Jobs 选项卡。该页展示了 Workflow 的列表，不但包括刚启动的 Workflow Jobs，还包括 Coordinator Job、Bundle Jobs、System Info、Instrumentation、Settings 选项卡。

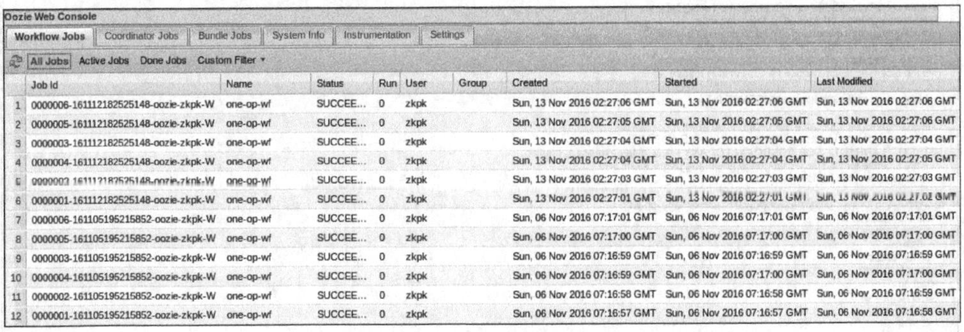

图 10-6　Oozie 的控制台界面

10.4.2　获取作业信息

用户可以使用 Oozie 控制台来观察工作流执行的进程和结果。当用户选定一个特定作业时，控制台会在页面上用 7 个选项卡（基本信息、定义、配置、日志、错误日志、审计日志、有向无环图）来展示该作业特有的信息。默认的 Job Info 展示作业的状态，以及所有已经启动的 Workflow 动作的当前状态，如图 10-7 所示。

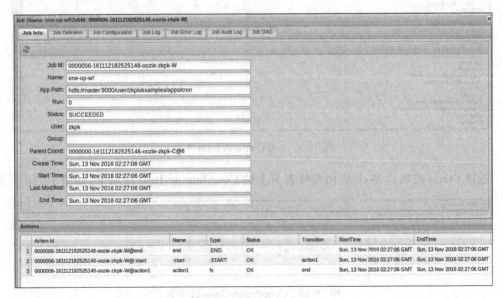

图 10-7　Job Info 界面

图 10-6 显示了和作业相关的信息，包括作业 ID、名字、应用程序路径、用户名等。它提供了作业的创建时间、名义时间、开始时间等。

切换到 Job Definition 可以看到作业定义，如图 10-8 所示。

切换到 Job Configuration，可以看到完整的作业配置，如图 10-9 所示。

之后显示的是作业日志、错误日志、审计日志、有向无环图，这里就不一一截图介绍了。

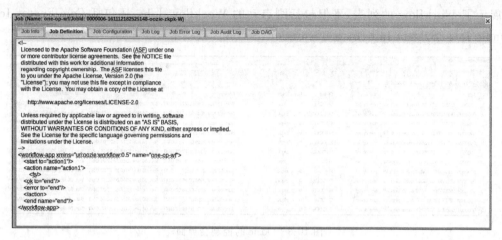

图 10-8　Job Definition 界面

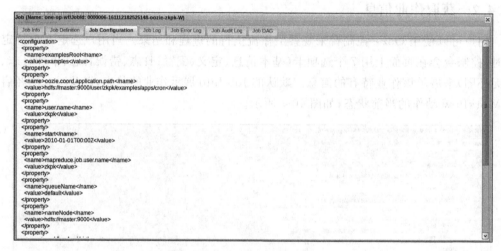

图 10-9　Job Configuration 界面

选择 Oozie 控制台 Web 应用程序首页上的 Coordinator Jobs 选项卡，如图 10-10 所示。

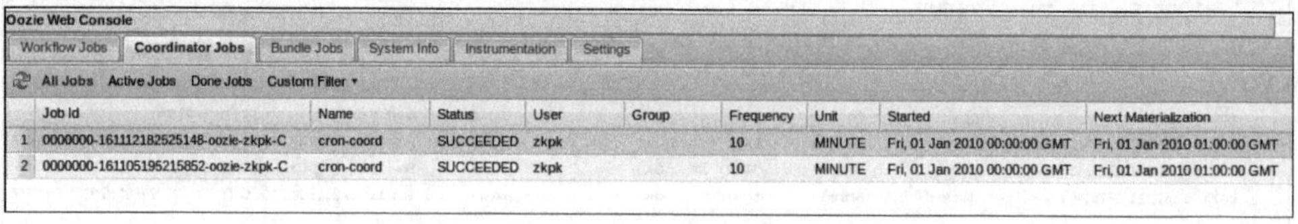

图 10-10　Coordinator Jobs 选项卡

选择其中一个 Coordinator 作业，展示所有 Coordinator 动作的信息，显示动作的 ID 字段，包含了显示动作启动顺序的前缀，对每个 Coordinator 动作而言，Oozie 控制台显示了对应 Workflow 动作的 ID（Exit ID 即第三列）。例如，对于动作 ID 为 0000000-161112182525148-oozie-zkpk-C 的 Coordinator 动作，对应 Workflow 作业的 ID 为 0000006-161112182525148-oozie-zkpk-W，如图 10-11 所示。

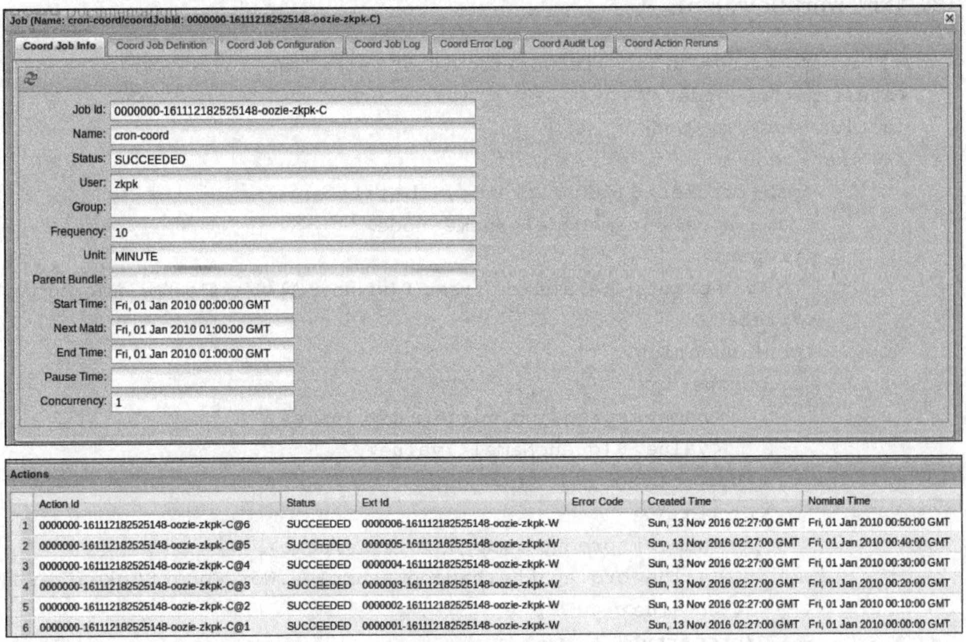

图 10-11 Coordinator 动作 ID 举例

10.5 Oozie 的高级特性

10.5.1 自定义 Oozie Workflow

本小节介绍一个通过 Oozie 来运行 MapReduce 中的 WordCount 的示例。使用数据为 Hadoop 安装目录下的 README.txt，并上传到 hdfs://master:50070/user/zkpk/README.txt。

配置环境变量。

export OOZIE_URL=http://master:11000/oozie

在 HDFS 中创建 workflow/mr_demo/wf。

[zkpk@master cron]$hdfs dfs -mkdir workflow/mr_demo/wf

（1）修改 job.properties。

oozie.wf.application.path=hdfs://master:9000/user/zkpk/workflow/mr_demo/wf
#Hadoop"R
jobTracker=master:9001
#Hadoop"fs.default.name
nameNode=hdfs://master:9000/
#Hadoop"mapred.queue.name
queueName=default

(2) 修改 workflow.xml。

```xml
<workflow-app xmlns="uri:oozie:workflow:0.5" name="map-reduce-wf">
    <start to="mr-node"/>
    <action name="mr-node">
        <map-reduce>
            <job-tracker>${jobTracker}</job-tracker>
            <name-node>${nameNode}</name-node>
            <prepare>
                <delete path="${nameNode}/user/${wf:user()}/workflow/mr_demo/output"/>
            </prepare>
            <configuration>
                <property>
                    <name>mapred.job.queue.name</name>
                    <value>${queueName}</value>
                </property>
                <property>
                    <name>mapreduce.mapper.class</name>
                    <value>org.apache.hadoop.examples.WordCount$TokenizerMapper
                    </value>
                </property>
                <property>
                    <name>mapreduce.reducer.class</name>
                    <value>org.apache.hadoop.examples.WordCount$IntSumReducer</value>
                </property>
                <property>
                    <name>mapred.map.tasks</name>
                    <value>1</value>
                </property>
                <property>
                    <name>mapred.input.dir</name>
                    <value>/user/${wf:user()}/README.txt</value>
                </property>
                <property>
                    <name>mapred.output.dir</name>
                    <value>/user/${wf:user()}/workflow/mr_demo/output</value>
                </property>
            </configuration>
        </map-reduce>
        <ok to="end"/>
        <error to="fail"/>
    </action>
    <kill name="fail">
        <message>Map/Reduce failed, error message[${wf:errorMessage(wf:
        lastErrorNode())}]</message>
    </kill>
    <end name="end"/>
</workflow-app>
```

(3) 复制 workflow.xml 文件到 HDFS 上。

[zkpk@master cron]$hdfs dfs -put workflow.xml workflow/mr_demo/wf/

(4) 提交作业。

[zkpk@master oozie-4.2.0]$bin/oozie job -oozie http://master:11000/oozie -config /home/zkpk/oozie-4.2.0/examples/target/oozie-examples-4.2.0-examples/examples/apps/cron/job.properties -run

```
SLF4J: Class path contains multiple SLF4J bindings.
SLF4J: Found binding in [jar:file:/home/zkpk/oozie-4.2.0/distro/target/oozie-4.2
.0-distro/oozie-4.2.0/lib/slf4j-log4j12-1.6.6.jar!/org/slf4j/impl/StaticLoggerBi
nder.class]
SLF4J: Found binding in [jar:file:/home/zkpk/oozie-4.2.0/distro/target/oozie-4.2
.0-distro/oozie-4.2.0/lib/slf4j-simple-1.6.6.jar!/org/slf4j/impl/StaticLoggerBin
der.class]
SLF4J: Found binding in [jar:file:/home/zkpk/oozie-4.2.0/distro/target/oozie-4.2
.0-distro/oozie-4.2.0/libext/slf4j-log4j12-1.7.5.jar!/org/slf4j/impl/StaticLogge
rBinder.class]
SLF4J: See http://www.slf4j.org/codes.html#multiple_bindings for an explanation.
SLF4J: Actual binding is of type [org.slf4j.impl.Log4jLoggerFactory]
job: 0000031-161112182525148-oozie-zkpk-W
```

(5) 使用 bin/oozie job 0000031-161112182525148-oozie-zkpk-W 即可查看流程状态。

[zkpk@master oozie-4.2.0]$bin/oozie job -info 0000031-161112182525148-oozie-zkpk-W

```
Job ID : 0000031-161112182525148-oozie-zkpk-W
------------------------------------------------------------------------------
Workflow Name : map-reduce-wf
App Path      : hdfs://master:9000/user/zkpk/workflow/mr_demo/wf
Status        : RUNNING
Run           : 0
User          : zkpk
Group         : -
Created       : 2016-11-13 09:57 GMT
Started       : 2016-11-13 09:57 GMT
Last Modified : 2016-11-13 09:57 GMT
Ended         : -
CoordAction ID: -

Actions
------------------------------------------------------------------------------
ID                                                                        St
atus     Ext ID             Ext Status  Err Code
------------------------------------------------------------------------------
0000031-161112182525148-oozie-zkpk-W@mr-node                              PR
EP        -                  -           -
------------------------------------------------------------------------------
0000031-161112182525148-oozie-zkpk-W@:start:                              OK
          -                  OK          -
------------------------------------------------------------------------------
```

(6) 流程结束后,查看流程状态,并在对应的目录查看输出结果。

10.5.2 使用 Oozie JavaAPI

用 Java 的 API 来提交代码,代码是在物理机上写的,然后提交到集群中去。要在任意一台机器上向 Oozie 提交作业,需要对 Hadoop 的 core-site.xml 文件进行设置,复制到所有机器上，然后重启 Hadoop 集群。

```xml
<property>
    <name>hadoop.proxyuser.zkpk.hosts</name>
    <value>*</value>
</property>
<property>
    <name>hadoop.proxyuser.zkpk.groups</name>
    <value>*</value>
</property>
```

这里都设置成星号，则可以是任意机器、任意账号。zkpk 是虚拟机账户。

在 IDE 中建立 Demo 类，并导入 Oozie 的 jar 包。

```
[zkpk@master ~]$ cd
oozie-4.2.0/distro/target/oozie-4.2.0-distro/oozie-4.2.0
[zkpk@master oozie-4.2.0]$ tar -zxvf oozie-client-4.2.0.tar.gz
[zkpk@master oozie-4.2.0]$ cd oozie-client-4.2.0/lib
```

jar 包如下所示（下列代码加两个 jar 包就可以，但为了以后写更多的类，可以全添加进去）。

```
[zkpk@master lib]$ ls
activemq-client-5.8.0.jar
commons-beanutils-1.7.0.jar
commons-beanutils-core-1.8.0.jar
commons-cli-1.2.jar
commons-codec-1.4.jar
commons-collections-3.2.1.jar
commons-configuration-1.6.jar
commons-digester-1.8.jar
commons-lang-2.4.jar
commons-logging-1.1.jar
geronimo-j2ee-management_1.1_spec-1.0.1.jar
geronimo-jms_1.1_spec-1.1.1.jar
guava-11.0.2.jar
hadoop-core-1.2.1.jar
hawtbuf-1.9.jar
jackson-core-asl-1.8.8.jar
jackson-mapper-asl-1.8.8.jar
json-simple-1.1.jar
jsr305-1.3.9.jar
log4j-1.2.16.jar
oozie-client-4.2.0.jar
oozie-hadoop-auth-hadoop-1-4.2.0.jar
slf4j-api-1.6.6.jar
slf4j-simple-1.6.6.jar
xercesImpl-2.10.0.jar
xml-apis-1.4.01.jar
```

代码非常简单，先创建一个 OozieClient 对象，再创建一个配置文件 Properties 类，然后把文件的 Job.Properties 里写的所有参数都设置进去就行了，然后调用 run 方法。

```java
import java.util.Properties;
import org.apache.oozie.client.OozieClient;
import org.apache.oozie.client.OozieClientException;

public class Demo
```

```
{
    public static void main(String[] args)
    {
        OozieClient wc=new OozieClient("http://192.168.111.128:11000/oozie");
        Properties conf=wc.createConfiguration();

        conf.setProperty("nameNode", "hdfs://192.168.111.128:9000");
        conf.setProperty("queueName", "default");
        conf.setProperty("examplesRoot", "examples");
        conf.setProperty("oozie.wf.application.path", "${nameNode}/user/zkpk/
        ${examplesRoot}/apps/map-reduce");
        conf.setProperty("outputDir", "map-reduce");
        conf.setProperty("jobTracker", "http://192.168.111.128:9001");
        try
        {
            String jobId=wc.run(conf);
        } catch(OozieClientException e)
        {
            //TODO Auto-generated catch block
            e.printStackTrace();
        }

    }
}
```

在控制台界面查看,发现提交成功,如图 10-12 所示。

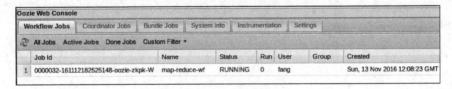

图 10-12 提交成功

本章小结

本章介绍了工作流调度组件 Oozie,及其安装及相关组件的功能与使用,并给出了简单的程序运行案例。学习本章之后要掌握：Oozie 的安装及配置；Workflow、Coordinator、Bundle 组件的功能及具体使用方法；能编写相关程序,并进行 Oozie 任务的部署与执行。

习 题

(1) Oozie 的作用是什么？在 Hadoop 生态系统中处于什么位置？

(2) Oozie 的组件有哪些？各有什么功能？

(3) 编写并部署 Hive 类任务。Hive 动作的语法格式如下所示。

```xml
<workflow-app name="[WF-DEF-NAME]" xmlns="uri:oozie:workflow:0.2">
    ...
    <action name="[NODE-NAME]">
        <hive xmlns="uri:oozie:hive-action:0.2">
            <job-tracker>[JOB-TRACKER]</job-tracker>
            <name-node>[NAME-NODE]</name-node>
            <prepare>
                <delete path="[PATH]" />
                ...
                <mkdir path="[PATH]" />
                ...
            </prepare>
            <configuration>
                <property>
                    <name>[PROPERTY-NAME]</name>
                    <value>[PROPERTY-VALUE]</value>
                </property>
                ...
            </configuration>
            <script>[HIVE-SCRIPT]</script>
            <param>[PARAM-VALUE]</param>
            ...
        </hive>
        <ok to="[NODE-NAME]" />
        <error to="[NODE-NAME]" />
    </action>
    ...
</workflow-app>
```

第11章

离线计算实例

本章摘要

本书前面的章节对大数据离线分析的工具进行了详细的介绍,那么在实际应用中这些技术是如何整合在一起的?本章将通过两个实际的案例对这些工具的使用进行介绍,两个案例中将分别使用 Hive 和 Pig 对数据进行分析。希望通过本章的介绍,使读者能够对大数据离线分析框架有更深入的认识。

11.1 微博历史数据分析

对微博进行数据分析,有利于人们更好地去做微博。那么,具体收集哪些微博数据?从数据上看又能看出什么?

(1) 粉丝:从粉丝来看,粉丝数越多的人自然越能引起人注意。那么粉丝数增长快的人又能说明什么问题?

(2) 内容:从博主的微博内容来看,都是什么类型的?是单纯的原创,还是活动类的(如投票,有奖转发)?博主每天发内容的频率如何?微博内容的来源是原创的产品信息,还是各类的分享,抑或是来自 PP 内容库?

(3) 转发:从微博转发来看,什么样的微博转发高?转发数多少?在转发的同时评论的人多吗?如果某微博转发多,而且评论的人多,能说明什么问题?转发高的微博的内容是什么类型,为什么转发高?还有是否可以私信?企业认证版的微博板块上有什么不同?

(4) 关注:从关注来看,博主都关注了哪些人,什么行业的,是否是同行业?关注的人里,加 V 认证的人多吗?如果加 V 认证的人多,能说明什么问题?

经过对微博中上述历史数据的收集后不难看出,博主提高微博影响力的最主要的策略是提高微博的转发数。那么怎样提高微博的转发数呢?最重要的还是微博的内容。从数据来看,往往活动内容的转发数会非常高,一般都会有几百。例如,转发词条微博并且@3位好友,就有机会获得奖品。更甚者,博主会将某一条微博置顶,那么这条微博的转发数自然而然就会提高了。

11.1.1 数据结构

本小节中使用的数据是新浪微博的日志文件,数据以 json 格式存储,其中数据的字段和含义见表 11-1。

表 11-1　微博数据字段名及含义

字 段 名	含 义
beCommentWeiboId	是否是评论微博(ID)
beForwardWeiboId	是否是转发微博(ID)
catchTime	抓取时间
commentCount	评论人数
content	内容
createTime	创建时间
info1	信息 1
info2	信息 2
info3	信息 3
mlevel	用户等级
musicurl	音乐链接
pic_list	照片列表(可以有多个)
praiseCount	点赞人数
reportCount	转发人数
source	数据来源
userId	用户 ID
videourl	视频链接
weiboId	微博 ID
weiboUrl	微博网址

11.1.2　需求分析

(1) 获取微博被转发数最多的前 n 位用户的 ID。

用途：可以用来做关注推荐。

(2) 分别获取评论数量、转发数量、点赞数量最多的前 n 条微博。

用途：可以分析用户行为。例如，当某个用户评论了某条微博，说明用户比较能对此类微博信息产生共鸣；转发某条微博，说明用户比较愿意跟其他用户分享此类信息。

(3) 根据使用终端品牌对用户进行分类。

用途：可以对产品广告或者 APP 广告实现精准投放。例如，购买过小米 5 的用户很有可能会购买小米 6。

(4) 获取评论数量、转发数量、点赞数量之和最多的前 n 条原创微博。

用途：找出微博中关注程度比较高的前几条，也可以更有针对性地对某些比较活跃的用户进行统计分析。

11.1.3　需求实现

1. 数据导入

将本地的数据上传到 HDFS 时分两种情况：一种是数据以文本形式存储在本地磁盘

中;另一种是数据以数据库文件形式存储在数据库(如 MySQL)中。这里针对第一种情况进行演示。

(1) 数据以文本文件的形式存储在本地磁盘中。例如,数据保存在~/resources/data/weibo/目录下。

首先在 HDFS 中创建一个文件夹,用于在 HDFS 文件系统中存储微博数据。

[zkpk@master ~]$hdfs dfs -mkdir -p workflow/demo/weibo/data

此时使用下列语句,将本地数据上传到 HDFS 文件系统中。

[zkpk@master ~]$hdfs dfs -put resources/data/weibo/* workflow/demo/weibo/data

输入查看指令可以查看已上传的数据。

[zkpk@master ~]$hdfs dfs -ls workflow/demo/weibo/data

部分结果如图 11-1 所示。

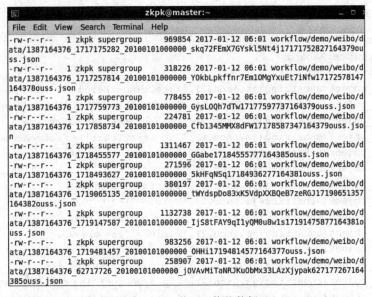

图 11-1　查看已上传的数据

(2) 如果数据存储在 MySQL 中,则可以使用 Sqoop 将数据从 MySQL 导入 Hive。

2. 数据预处理

使用 Oozie 工作流组件进行数据表的创建和视图的创建时,首先要修改和创建一些脚本文件,然后将这些脚本文件上传到 HDFS 中。

(1) 切换 zkpk 用户。

[zkpk@master ~]$su zkpk

输入密码:

zkpk

将 hive-site.xml 文件上传到 HDFS 中。

```
[zkpk@master ~]$hdfs dfs -put
~/apache-hive-0.13.1-bin/conf/hive-site.xml workflow/demo/weibo
```

(2) 使用 gedit 命令来创建 create.hql、createview.hql、workflow.xml、hive-default.xml、job.properties 这 5 个文件。例如：

```
[zkpk@master ~]$gedit ~/create.hql
```

这 5 个文件的内容如下。

① create.hql。

```
use weibo;
create external table IF NOT EXISTS data
(json string)
row format delimited lines terminated by "\n"
stored as textfile
location "/user/zkpk/workflow/demo/weibo/data";
```

② createview.hql。

```
use weibo;
create view userRecord
as select
get_json_object(substring(js.json,2),'$.beCommentWeiboId')asbeCommentWeiboId,
get_json_object(substring(js.json,2),'$.beForwardWeiboId')asbeForwardWeiboId,
get_json_object(substring(js.json,2),'$.catchTime')as catchTime,
get_json_object(substring(js.json,2),'$.commentCount')as commentCount,
get_json_object(substring(js.json,2),'$.content')as content,
get_json_object(substring(js.json,2),'$.createTime')as createTime,
get_json_object(substring(js.json,2),'$.info1')as info1,
get_json_object(substring(js.json,2),'$.info2')as info2,
get_json_object(substring(js.json,2),'$.info3')as info3,
get_json_object(substring(js.json,2),'$.mlevel')as mlevel,
get_json_object(substring(js.json,2),'$.musicurl')as musicurl,
get_json_object(substring(js.json,2),'$.pic_list')as pic_list,
get_json_object(substring(js.json,2),'$.praiseCount')as praiseCount,
get_json_object(substring(js.json,2),'$.reportCount')as reportCount,
get_json_object(substring(js.json,2),'$.source')as source,
get_json_object(substring(js.json,2),'$.userId')as userId,
get_json_object(substring(js.json,2),'$.videourl')as videourl,
get_json_object(substring(js.json,2),'$.weiboId')as weiboId,
get_json_object(substring(js.json,2),'$.weiboUrl')as weiboUrl
from(select json from data)js;
```

③ workflow.xml。

```
<workflow-app name="weibo_wf" xmlns="uri:oozie:workflow:0.5">
    <start to="table_create" />
    <action name="table_create">
        <hive xmlns="uri:oozie:hive-action:0.5">
            <job-tracker>${jobTracker}</job-tracker>
            <name-node>${nameNode}</name-node>
```

```xml
            <job-xml>hive-site.xml</job-xml>
            <configuration>
                <property>
                    <name>mapred.job.queue.name</name>
                    <value>${queueName}</value>
                </property>
            </configuration>
            <script>create.hql</script>
        </hive>
        <ok to="view_create" />
        <error to="fail" />
    </action>

    <action name="view_create">
        <hive xmlns="uri:oozie:hive-action:0.5">
            <job-tracker>${jobTracker}</job-tracker>
            <name-node>${nameNode}</name-node>
            <job-xml>hive-site.xml</job-xml>
            <configuration>
                <property>
                    <name>mapred.job.queue.name</name>
                    <value>${queueName}</value>
                </property>
            </configuration>
            <script>createview.hql</script>
        </hive>
        <ok to="end" />
        <error to="fail" />
    </action>
    <kill name="fail">
        <message>error message[${wf:errorMessage(wf:lastErrorNode())}]</message>
    </kill>
    <end name="end" />
</workflow-app>
```

④ hive-default.xml。

```xml
<?xml version="1.0"?>
<?xml-stylesheet type="text/xsl" href="configuration.xsl"?>
<configuration>
    <property>
        <name>hive.metastore.warehouse.dir</name>
        <value>/user/hive/warehouse</value>
    </property>
    <property>
        <name>hive.metastore.uris</name>
        <value>thrift://master:9083</value>
    </property>
</configuration>
```

⑤ job.properties。

```
nameNode=hdfs://master:9000
jobTracker=master:18040
queueName=default
examplesRoot=examples
oozieRoot=oozie
hiveRoot=hive

oozie.use.system.libpath=true
oozie.libpath=share/lib/lib_20170101021854/hive
oozie.action.sharelib.for.hive=hive,hcatalog,sqoop
oozie.wf.application.path=${nameNode}/user/zkpk/workflow/demo/weibo
oozie.hive.defaults=${nameNode}/user/zkpk/workflow/demo/weibo/hive-default.xml
```

将前 4 个脚本文件上传到 HDFS 中。

[zkpk@master ~]$ hdfs dfs -put create.hql createview.hql workflow.xml hive-default.xml workflow/demo/weibo

然后使用 ls 命令查看：

```
[zkpk@master ~]$ hdfs dfs -ls workflow/demo/weibo
17/01/24 02:16:25 WARN util.NativeCodeLoader: Unable to load native-hadoop librar
y for your platform... using builtin-java classes where applicable
Found 6 items
-rw-r--r--   1 zkpk supergroup        182 2017-01-16 08:02 workflow/demo/weibo/cr
eate.hql
-rw-r--r--   1 zkpk supergroup       1472 2017-01-18 07:22 workflow/demo/weibo/cr
eateview.hql
drwxr-xr-x   - zkpk supergroup          0 2017-01-12 06:01 workflow/demo/weibo/da
ta
-rw-r--r--   1 zkpk supergroup        447 2017-01-24 00:50 workflow/demo/weibo/hi
ve-default.xml
-rw-r--r--   1 zkpk supergroup        670 2017-01-24 00:57 workflow/demo/weibo/hi
ve-site.xml
-rw-r--r--   1 zkpk supergroup       1467 2017-01-18 08:02 workflow/demo/weibo/wo
rkflow.xml
```

（3）开启相关服务。

在执行 Oozie 作业之前，需要启动 historyserver。

首先使用 gedit 修改 mapred-site.xml。

[zkpk@master ~]$gedit ~/hadoop-2.5.1/etc/hadoop/mapred-site.xml

在<property>标签中添加如下属性。

```
<name>mapreduce.jobhistory.address</name>
<value>master:10020</value>
<name>mapreduce.jobhistory.webapp.address</name>
<value>master:19888</value>
```

完成后需要启动 historyserver 服务。Hadoop 启动 jobhistoryserver 来实现 Web 查看作业的历史运行情况。由于在启动 HDFS 和 Yarn 进程之后，jobhistoryserver 进程并没有启动，需要手动启动，因此运行下面的指令启动 historyserver。

```
[zkpk@master ~]$ $HADOOP_HOME/sbin/mr-jobhistory-daemon.sh start historyserver
starting historyserver, logging to /home/zkpk/hadoop-2.5.1/logs/mapred-zkpk-histo
ryserver-master.out
```

要在装有 Hive 的主节点上启动 Hive 的 metastore 服务。

另外开启一个 terminal 终端,启动 metastore 服务,如图 11-2 所示。

```
[zkpk@master Desktop]$ hive --service metastore
```

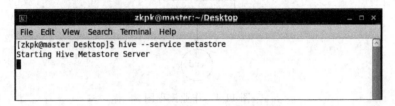

图 11-2　启动 metastore 服务

使用期间不要将其停止或关闭。

(4) 运行 Oozie 作业。

```
[zkpk@master oozie-4.2.0]$ cd
~/oozie-4.2.0/distro/target/oozie-4.2.0-distro/oozie-4.2.0
[zkpk@master oozie-4.2.0]$ bin/oozie job -oozie http://master:11000/oozie -config
/home/zkpk/job.properties -run
SLF4J: Class path contains multiple SLF4J bindings.
SLF4J: Found binding in [jar:file:/home/zkpk/oozie-4.2.0/distro/target/oozie-4.2
.0-distro/oozie-4.2.0/lib/slf4j-log4j12-1.6.6.jar!/org/slf4j/impl/StaticLoggerBi
nder.class]
SLF4J: Found binding in [jar:file:/home/zkpk/oozie-4.2.0/distro/target/oozie-4.2
.0-distro/oozie-4.2.0/lib/slf4j-simple-1.6.6.jar!/org/slf4j/impl/StaticLoggerBin
der.class]
SLF4J: Found binding in [jar:file:/home/zkpk/oozie-4.2.0/distro/target/oozie-4.2
.0-distro/oozie-4.2.0/libext/slf4j-log4j12-1.7.5.jar!/org/slf4j/impl/StaticLogge
rBinder.class]
SLF4J: See http://www.slf4j.org/codes.html#multiple_bindings for an explanation.
SLF4J: Actual binding is of type [org.slf4j.impl.Log4jLoggerFactory]
job: 0000015-170101020026374-oozie-zkpk-W
[zkpk@master oozie-4.2.0]$
```

可以进入默认端口 11000,进入网页界面查看结果,如图 11-3 所示。

图 11-3　网页界面查看结果

作业的流程图如图 11-4 所示。

创建完表和视图后,可以使用 Hive 执行一个查询进行测试。

```
hive>use weibo;
hive>select count(*)from data;
```

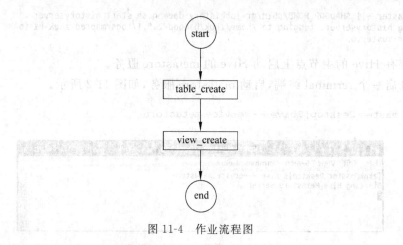

图 11-4 作业流程图

```
MapReduce Total cumulative CPU time: 23 seconds 990 msec
Ended Job = job_1483265868973_0013
MapReduce Jobs Launched:
Job 0: Map: 4  Reduce: 1   Cumulative CPU: 23.99 sec   HDFS Read: 918515577 HDFS
 Write: 8 SUCCESS
Total MapReduce CPU Time Spent: 23 seconds 990 msec
OK
1451868
Time taken: 115.665 seconds, Fetched: 1 row(s)
```

3. 数据需求实现

(1) 查询被转发数最多的前 3 位用户，返回用户 ID 和被转发数。

使用 Oozie 工作流进行实现。首先通过 Hive 动作执行查询，并将返回结果存储到 Hive 表中。然后使用 sqoop 动作将 Hive 中的查询结果导入 MySQL 中。最后，在 MySQL 中查看结果并使用。

在运行 Oozie 工作流之前，需要做一些准备工作。

在 HDFS 文件系统中创建一个用于此工作流的文件夹。

```
[zkpk@master ~]$hdfs dfs -mkdir -p workflow/demo/weibo/query1
```

将用到的配置文件及脚本文件上传到此文件夹下：

```
[zkpk@master ~]$hdfs dfs -put query1/* workflow/demo/weibo/query1
```

① 文件 workflow.xml 的内容如下：

```xml
<workflow-app name="weibo_wf" xmlns="uri:oozie:workflow:0.5">
    <start to="create" />
    <action name="create">
        <hive xmlns="uri:oozie:hive-action:0.5">
            <job-tracker>${jobTracker}</job-tracker>
            <name-node>${nameNode}</name-node>
            <job-xml>hive-site.xml</job-xml>
            <configuration>
                <property>
                    <name>mapred.job.queue.name</name>
                    <value>${queueName}</value>
```

```
                </property>
            </configuration>
            <script>create1.hql</script>
        </hive>
        <ok to="query" />
        <error to="fail" />
    </action>
    <action name="query">
        <hive xmlns="uri:oozie:hive-action:0.5">
            <job-tracker>${jobTracker}</job-tracker>
            <name-node>${nameNode}</name-node>
            <job-xml>hive-site.xml</job-xml>
            <configuration>
                <property>
                    <name>mapred.job.queue.name</name>
                    <value>${queueName}</value>
                </property>
            </configuration>
            <script>query1.hql</script>
        </hive>
        <ok to="export" />
        <error to="fail" />
    </action>
    <action name="export">
        <sqoop xmlns="uri:oozie:sqoop-action:0.2">
            <job-tracker>${jobTracker}</job-tracker>
            <name-node>${nameNode}</name-node>
            <configuration>
                <property>
                    <name>mapred.job.queue.name</name>
                    <value>${queueName}</value>
                </property>
            </configuration>
            <command>export --connect jdbc:mysql://master:3306/test --username
            hadoop --password hadoop --table query1 --input-fields-terminated-
            by '\001' --export-dir /user/zkpk/workflow/demo/weibo/query1/data
            </command>
        </sqoop>
        <ok to="end" />
        <error to="fail" />
    </action>
    <kill name="fail">
        <message>error
            message[${wf:errorMessage(wf:lastErrorNode())}]</message>
    </kill>
    <end name="end" />
</workflow-app>
```

② 文件 create1.hql 的内容如下：

use weibo;

```
create external table if not exists query1(id string,cnt bigint) location "/user/
zkpk/workflow/demo/weibo/query1/data";
```

③ 文件 query1.hql 的内容如下：

```
use weibo;
insert overwrite table query1 select userId,sum(reportCount) as cnt from userRecord
group by userId order by cnt DESC limit 3;
```

④ 文件 job.properties 的内容如下：

```
nameNode=hdfs://master:9000
jobTracker=master:18040
queueName=default
examplesRoot=examples
oozieRoot=oozie
hiveRoot=hive
oozie.use.system.libpath=true
oozie.libpath=share/lib/lib_20170101021854/hive
oozie.action.sharelib.for.hive=hive,hcatalog,sqoop
oozie.wf.application.path=${nameNode}/user/zkpk/workflow/demo/weibo/query1
oozie.hive.defaults=${nameNode}/user/zkpk/workflow/demo/weibo/hive-default.xml
```

文件上传完毕后，使用 Oozie 命令执行工作流。

```
[zkpk@master oozie-4.2.0]$bin/oozie job -oozie
http://master:11000/oozie -config
/home/zkpk/query1/job.properties -run
```

等待运行完成后，可以在 MySQL 中查看结果。

```
mysql> select * from query1;
+------------+----------+
| id         | cnt      |
+------------+----------+
| 1793285524 | 76454805 |
| 1629810574 | 73656898 |
| 2803301701 | 68176008 |
+------------+----------+
3 rows in set (0.00 sec)
```

(2) 分别获取评论数量、转发数量、点赞数量最多的前 5 条微博。

在运行 Oozie 工作流前，需要做一些准备工作。首先在 HDFS 文件系统中创建一个用于此工作流的文件夹，然后将用到的配置文件和脚本文件上传到此文件夹下。

```
[zkpk@master ~]$hdfs dfs -mkdir -p workflow/demo/weibo/query2
[zkpk@master ~]$hdfs dfs -put query2/* workflow/demo/weibo/query1
```

① 文件 workflow.xml 的内容如下：

```
<workflow-app name="weibo_wf" xmlns="uri:oozie:workflow:0.5">
    <start to="query" />
    <action name="query">
        <hive xmlns="uri:oozie:hive-action:0.5">
            <job-tracker>${jobTracker}</job-tracker>
```

```xml
            <name-node>${nameNode}</name-node>
            <job-xml>hive-site.xml</job-xml>
            <configuration>
                <property>
                    <name>mapred.job.queue.name</name>
                    <value>${queueName}</value>
                </property>
            </configuration>
            <script>query2.hql</script>
        </hive>
        <ok to="export" />
        <error to="fail" />
    </action>
    <action name="export">
        <sqoop xmlns="uri:oozie:sqoop-action:0.2">
            <job-tracker>${jobTracker}</job-tracker>
            <name-node>${nameNode}</name-node>
            <configuration>
                <property>
                    <name>mapred.job.queue.name</name>
                    <value>${queueName}</value>
                </property>
            </configuration>
            <command>
                export --connect jdbc:mysql://master:3306/test --username hadoop --password hadoop --table query2 --input-fields-terminated-by '\001' --export-dir /user/zkpk/workflow/demo/weibo/query2/data
            </command>
        </sqoop>
        <ok to="end" />
        <error to="fail" />
    </action>
    <kill name="fail">
        <message>
            error message[${wf:errorMessage(wf:lastErrorNode())}]
        </message>
    </kill>
    <end name="end" />
</workflow-app>
```

② 文件 query2.hql 的内容如下：

```
use weibo;
create external table if not exists query2(wid string,ccnt bigint,rcnt bigint,pcnt bigint) location "/user/zkpk/workflow/demo/weibo/query2/data";
insert overwrite table query2 select weiboId, max(commentCount) as cnt, max(reportCount),max(praiseCount) from userRecord group by weiboId order by cnt+0 DESC limit 5;
insert into table query2 select weiboId,max(commentCount),max(reportCount) as cnt, max(praiseCount) from userRecord group by weiboId order by cnt+0 DESC limit 5;
insert into table query2 select weiboId, max(commentCount), max(reportCount), max
```

(praiseCount)as cnt from userRecord group by weiboid order by cnt+0 DESC limit 5;

③ 文件 job.properties 的内容如下：

```
nameNode=hdfs://master:9000
jobTracker=master:18040
queueName=default
examplesRoot=examples
oozieRoot=oozie
hiveRoot=hive
oozie.use.system.libpath=true
oozie.libpath=share/lib/lib_20170101021854/hive
oozie.action.sharelib.for.hive=hive,hcatalog,sqoop
oozie.wf.application.path=${nameNode}/user/zkpk/workflow/demo/weibo/query2
oozie.hive.defaults=${nameNode}/user/zkpk/workflow/demo/weibo/hive-default.xml
```

文件上传完毕后，使用 Oozie 命令执行工作流。

```
[zkpk@master oozie-4.2.0]$bin/oozie job -oozie
http://master:11000/oozie -config
/home/zkpk/query2/job.properties  -run
```

等待运行完成之后，可以在 MySQL 中查看结果。

```
mysql> select * from query2;
+-------------------+---------+---------+---------+
| wid               | ccnt    | rcnt    | pcnt    |
+-------------------+---------+---------+---------+
| 3640589771845876  |   97814 |   87041 | 1088041 |
| 3638085553880751  |   53271 |  150372 |  577252 |
| 3643995177446910  |  447113 | 1016059 |  537629 |
| 3649095539317717  |  428040 |  975168 |  530859 |
| 3640945981719667  |   17890 |   15211 |  405903 |
| 3524887173723617  |  818362 | 2692012 |       1 |
| 3516682531774365  |    1507 | 1897144 |    1393 |
| 3445888074596137  |  893224 | 1655525 |     124 |
| 3651029885742911  |  365870 | 1433730 |       1 |
| 3651064933352422  |  321344 | 1377553 |       1 |
| 3542033115012943  | 3839918 |   29127 |       1 |
| 3542005970239999  | 1668571 |     910 |     499 |
| 3542030276046506  | 1609132 |    9589 |       1 |
| 3541938278094601  | 1462048 |    1529 |     363 |
| 3543138209124897  | 1274700 |   12004 |    1397 |
+-------------------+---------+---------+---------+
15 rows in set (0.00 sec)
```

（3）根据使用终端品牌对用户进行分类。

① 使用 Oozie 工作流，文件 workflow.xml 的内容如下：

```xml
<workflow-app name="weibo_wf" xmlns="uri:oozie:workflow:0.5">
    <start to="query" />
    <action name="query">
        <hive xmlns="uri:oozie:hive-action:0.5">
            <job-tracker>${jobTracker}</job-tracker>
            <name-node>${nameNode}</name-node>
            <job-xml>hive-site.xml</job-xml>
            <configuration>
                <property>
                    <name>mapred.job.queue.name</name>
```

```xml
                <value>${queueName}</value>
            </property>
        </configuration>
        <script>query3.hql</script>
    </hive>
    <ok to="export" />
    <error to="fail" />
</action>
<action name="export">
    <sqoop xmlns="uri:oozie:sqoop-action:0.2">
        <job-tracker>${jobTracker}</job-tracker>
        <name-node>${nameNode}</name-node>
        <configuration>
            <property>
                <name>mapred.job.queue.name</name>
                <value>${queueName}</value>
            </property>
        </configuration>
        <command>export --connect "jdbc:mysql://master:3306/test?useUnicode=true&amp;characterEncoding=utf-8" --username hadoop --password hadoop --table query3 --input-fields-terminated-by '\001' --export-dir /user/zkpk/workflow/demo/weibo/query3/data</command>
    </sqoop>
    <ok to="end" />
    <error to="fail" />
</action>
<kill name="fail">
    <message>
       error  message[${wf:errorMessage(wf:lastErrorNode())}]
    </message>
</kill>
<end name="end" />
</workflow-app>
```

② 文件 query3.hql 的内容如下：

```
use weibo;
create external table if not exists query3(source string, cnt bigint) location "/user/zkpk/workflow/demo/weibo/query3/data";
insert overwrite table query3 select source, count (distinct userId) as cnt from userRecord group by source;
```

③ 文件 job.properties 的内容如下：

```
nameNode=hdfs://master:9000
jobTracker=master:18040
queueName=default
examplesRoot=examples
oozieRoot=oozie
hiveRoot=hive
oozie.use.system.libpath=true
oozie.libpath=share/lib/lib_20170101021854/hive
```

```
oozie.action.sharelib.for.hive=hive,hcatalog,sqoop
oozie.wf.application.path=${nameNode}/user/zkpk/workflow/demo/weibo/query3
oozie.hive.defaults=${nameNode}/user/zkpk/workflow/demo/weibo/hive-default.xml
```

文件上传完毕后,使用 Oozie 命令执行工作流。

```
[zkpk@master oozie-4.2.0]$bin/oozie job -oozie
http://master:11000/oozie -config
/home/zkpk/query3/job.properties    -run
```

等待运行完成之后,可以在 MySQL 中查看结果。

```
mysql>select * from query3 where source!='' order by cnt desc limit 20;
mysql> select * from query3 where source !='' order by cnt desc limit 20;
+------------------------+------+
| source                 | cnt  |
+------------------------+------+
| 新浪微博               | 1114 |
| iPhone客户端           |  921 |
| 专业版微博             |  712 |
| iPad客户端             |  559 |
| 媒体版微博             |  518 |
| Android客户端          |  430 |
| 微博搜索               |  422 |
| 分享按钮               |  320 |
| 新浪微博手机版         |  304 |
| 新浪博客               |  291 |
| 360浏览器超速版        |  217 |
| 360安全浏览器          |  210 |
| 优酷网连接分享         |  191 |
| 新浪视频               |  161 |
| 皮皮时光机             |  154 |
| 搜狗高速浏览器         |  136 |
| Weico.iPhone           |  134 |
| 关联博客               |  132 |
| 新浪长微博             |  129 |
| iOS                    |  120 |
+------------------------+------+
20 rows in set (0.00 sec)
```

(4) 获取评论数量、转发数量、点赞数量之和最多的前 n 条原创微博。

① 文件 workflow.xml 的内容如下:

```xml
<workflow-app name="weibo_wf" xmlns="uri:oozie:workflow:0.5">
    <start to="query" />
    <action name="query">
        <hive xmlns="uri:oozie:hive-action:0.5">
            <job-tracker>${jobTracker}</job-tracker>
            <name-node>${nameNode}</name-node>
            <job-xml>hive-site.xml</job-xml>
            <configuration>
                <property>
                    <name>mapred.job.queue.name</name>
                    <value>${queueName}</value>
                </property>
            </configuration>
            <script>query4.hql</script>
        </hive>
        <ok to="export" />
        <error to="fail" />
    </action>
```

```xml
<action name="export">
    <sqoop xmlns="uri:oozie:sqoop-action:0.2">
        <job-tracker>${jobTracker}</job-tracker>
        <name-node>${nameNode}</name-node>
        <configuration>
            <property>
                <name>mapred.job.queue.name</name>
                <value>${queueName}</value>
            </property>
        </configuration>
        <command>export --connect jdbc:mysql://master:3306/test --username
        hadoop --password hadoop --table query4 --input-fields-terminated-
        by '\001' --export-dir /user/zkpk/workflow/demo/weibo/query4/data</command>
    </sqoop>
    <ok to="end" />
    <error to="fail" />
</action>
<kill name="fail">
    <message>
        error message[${wf:errorMessage(wf:lastErrorNode())}]
    </message>
</kill>
<end name="end" />
</workflow-app>
```

② 文件 query4.hql 的内容如下：

```
use weibo;
create external table if not exists query4(wid string,cnt bigint) location "/user/
zkpk/workflow/demo/weibo/query4/data";
insert overwrite table query4 select weiboid,(a+b+c)as cnt from(select weiboid,max
(commentCount) as a,max (reportCount) as b, max (praiseCount) as c from userRecord
group by weiboId)t order by cnt+0 desc limit 10;
```

③ 文件 job.properties 的内容如下：

```
nameNode=hdfs://master:9000
jobTracker=master:18040
queueName=default
examplesRoot=examples
oozieRoot=oozie
hiveRoot=hive
oozie.use.system.libpath=true
oozie.libpath=share/lib/lib_20170101021854/hive
oozie.action.sharelib.for.hive=hive,hcatalog,sqoop
oozie.wf.application.path=${nameNode}/user/zkpk/workflow/demo/weibo/query4
oozie.hive.defaults=${nameNode}/user/zkpk/workflow/demo/weibo/hive-default.xml
```

文件上传完毕，使用 Oozie 命令执行工作流。

```
[zkpk@master oozie-4.2.0]$bin/oozie job -oozie
```

```
http://master:11000/oozie -config
/home/zkpk/query4/job.properties  -run
```

等待运行完成之后,可以在 MySQL 中查看结果。

```
mysql> select * from query4 order by cnt desc;
+-----------------+---------+
| wid             | cnt     |
+-----------------+---------+
| 3542033115012943 | 3869046 |
| 3524887173723617 | 3510375 |
| 3445888074596137 | 2548873 |
| 3643995177446910 | 2000801 |
| 3649095539317717 | 1934067 |
| 3516682531774365 | 1900044 |
| 3651029885742911 | 1799601 |
| 3635801391335069 | 1756481 |
| 3646005440738168 | 1727637 |
| 3651064933352422 | 1698898 |
+-----------------+---------+
10 rows in set (0.00 sec)
```

11.2 电商销售数据分析

洞察消费者全部的行为,是商业社会一直以来的梦想。今天,这个梦想似乎不再遥不可及,因为我们手中已握有最重要的拼图,它就是数据。无论是银行,还是京东、1 号店等电子商务平台,数据所展现的其实都是消费者背后的真实生活场景。其中,以用户行为、产品解决方案和数据深度挖掘为基础交叉研究出的数据尤为有价值。对企业而言,最重要的是如何对不同来源、不同类型的数据进行建模和分析。这项工作虽然枯燥,但是不可或缺。

说到数据分析,大家可能就会想到回归、聚类等,不过对于电商的小伙伴来说,这些都太复杂了。在实际分析的时候,其实并不需要这么复杂的算法,大家需要的只是对比、细分、转化、分类,只要掌握了这四种思想,基本上已经可以应付日常的分析工作了。

电子商务信息系统最核心的能力是大数据能力,包括大数据处理、数据分析和数据挖掘能力。无论是电商平台(如淘宝)还是在电商平台上销售产品的卖家,都需要掌握大数据分析的能力。越成熟的电商平台,越需要通过大数据能力驱动电子商务运营的精细化,更好地提升运营效果,提升业绩。

对于消费者而言,可以通过数据进行购物决策。网上购物因为商品看得见、摸不着,用户只能透过店铺信誉、成交量、买家评价等信息来判断店家是否"靠谱"。同时,消费者网购家电等产品的过程中对产品的评价越来越敏感,所以真实可靠的用户评价数据能够给家电消费者提供决策参考依据。而大量评价数据的汇总,则在潜移默化中,成为量化电商平台品质和服务能力的重要参考依据。

除此之外,电商大数据的应用场景还将涉及物流仓储优化、个性化推荐、金融信用、风控等多个环节。

11.2.1 数据结构

本节使用的数据是天猫商城的日志文件,数据以 json 格式存储。数据的字段包括:UID、创建时间、商品名称、支付价(含邮费)、数量、商品 ID,详见表 11-2。

表 11-2　数据字段

字 段 名	含　　义	字 段 名	含　　义
uid	UID	price	支付价（含邮费）
time	创建时间	number	数量
pname	商品名称	pid	商品 ID

11.2.2　需求分析

1. 商品类分析

（1）销售额前 n 名的商品。

（2）订单数前 n 名的商品。

2. 网站销售类分析

（1）一段时间内的总销售额。

（2）一段时间内的总订单量。

11.2.3　需求实现

1. 商品类分析

（1）销售额前 n 名的商品

使用 Oozie 工作流进行实现。首先通过 Pig 动作，执行查询并返回结果存储到 HDFS 文件系统中。然后使用 Sqoop 动作，将存储的查询结果导入到 MySQL 中。最后在 MySQL 中查看结果并使用。

与前一节类似，在运行 Oozie 工作流之前，需要做一些准备工作：在 HDFS 文件系统中创建一个用于此工作流的文件夹。

```
[zkpk@master ~]$ hdfs dfs -mkdir -p tmall/q1
```

将用到的配置文件及脚本文件上传到此文件夹下：

```
[zkpk@master ~]$ hdfs dfs -put tmall/q1/*  tmall/q1
```

workflow.xml 文件内容如下：

```
<workflow-app name="weibo_wf" xmlns="uri:oozie:workflow:0.5">
    <start to="query" />
    <action name="query">
        <pig>
            <job-tracker>${jobTracker}</job-tracker>
            <name-node>${nameNode}</name-node>
            <configuration>
                <property>
                    <name>oozie.launcher.mapred.child.java.opts</name>
                    <value>-Xmx2048m</value>
                </property>
                <property>
                    <name>pig.spill.extragc.size.threshold</name>
```

```xml
                <value>100000000</value>
            </property>
            <property>
                <name>mapred.child.java.opts</name>
                <value>-Xmx2048m</value>
            </property>
            <property>
                <name>mapred.user.jobconf.limit</name>
                <value>100000000</value>
            </property>
        </configuration>
        <script>q1.pig</script>
    </pig>
    <ok to="export" />
    <error to="fail" />
</action>
 <action name="export">
    <sqoop xmlns="uri:oozie:sqoop-action:0.2">
        <job-tracker>${jobTracker}</job-tracker>
        <name-node>${nameNode}</name-node>
        <configuration>
            <property>
                <name>mapred.job.queue.name</name>
                <value>${queueName}</value>
            </property>
        </configuration>
        <command>export - - connect " jdbc: mysql://master: 3306/tmall" - -
        username hadoop - - password hadoop - - table r1 - - input - fields -
        terminated-by '\t' --export-dir /user/zkpk/tmall/r1</command>
    </sqoop>
    <ok to="end" />
    <error to="fail" />
</action>
<kill name="fail">
    <message>error
        message[${wf:errorMessage(wf:lastErrorNode())}]</message>
</kill>
<end name="end" />
</workflow-app>
```

q1.pig 文件内容如下：

```
data = load '/user/zkpk/tmall/data/tmall - 201412.csv' as (uid, time, pname, price,
number,pid);
A=group data by pid;
B=foreach A generate group,SUM(data.number);
C=order B by $1 desc;
D=limit C 10;
store D into '/user/zkpk/tmall/r1';
```

job.properties 文件内容如下：

```
nameNode=hdfs://master:9000
jobTracker=master:18040
queueName=default
examplesRoot=examples
oozieRoot=oozie

oozie.use.system.libpath=true
oozie.libpath=share/lib/lib_20170101021854/pig
oozie.wf.application.path=${nameNode}/user/zkpk/tmall/q1
```

文件上传完毕后,使用 oozie 命令执行工作流。

```
[zkpk@master oozie-4.2.0]$bin/oozie job -oozie
http://master:11000/oozie -config
/home/zkpk/tmall/q1/job.properties  -run
```

等待运行完成之后,可以在 MySQL 中查看结果。

```
mysql> select * from r1 order by count+0 desc;
+-------------+--------+
| pid         | count  |
+-------------+--------+
| 17359010576 | 78046.0 |
| 17525034357 | 62294.0 |
| 35570005627 | 14703.0 |
| 38330727924 | 8703.0  |
| 37961093174 | 8659.0  |
| 36826954133 | 7639.0  |
| 37960797318 | 7485.0  |
| 17655706452 | 6399.0  |
| 27529496581 | 4892.0  |
| 24965336234 | 4838.0  |
+-------------+--------+
10 rows in set (0.07 sec)
```

(2) 订单数前 n 名的商品

使用 Oozie 工作流,workflow.xml 文件内容如下:

```xml
<workflow-app name="weibo_wf" xmlns="uri:oozie:workflow:0.5">
    <start to="query" />
    <action name="query">
        <pig>
            <job-tracker>${jobTracker}</job-tracker>
            <name-node>${nameNode}</name-node>
            <configuration>
                <property>
                    <name>oozie.launcher.mapred.child.java.opts</name>
                    <value>-Xmx2048m</value>
                </property>
                <property>
                    <name>pig.spill.extragc.size.threshold</name>
                    <value>100000000</value>
                </property>
                <property>
                    <name>mapred.child.java.opts</name>
                    <value>-Xmx2048m</value>
                </property>
```

```xml
                    <property>
                        <name>mapred.user.jobconf.limit</name>
                        <value>100000000</value>
                    </property>
                </configuration>
                <script>q2.pig</script>
            </pig>
            <ok to="export" />
            <error to="fail" />
        </action>
        <action name="export">
            <sqoop xmlns="uri:oozie:sqoop-action:0.2">
                <job-tracker>${jobTracker}</job-tracker>
                <name-node>${nameNode}</name-node>
                <configuration>
                    <property>
                        <name>mapred.job.queue.name</name>
                        <value>${queueName}</value>
                    </property>
                </configuration>
                <command>export --connect "jdbc:mysql://master:3306/tmall" --
                username hadoop --password hadoop --table r2 --input-fields-
                terminated-by '\t' --export-dir /user/zkpk/tmall/r2</command>
            </sqoop>
            <ok to="end" />
            <error to="fail" />
        </action>
        <kill name="fail">
            <message>error
                message[${wf:errorMessage(wf:lastErrorNode())}]</message>
        </kill>
        <end name="end" />
</workflow-app>
```

q2.pig 文件内容如下:

```
data=load '/user/zkpk/tmall/data/tmall-201412.csv' as (uid,time,pname,price,number,pid);
A=group data by pid;
B=foreach A generate group,COUNT(data);
C=order B by $1 desc;
D=limit C 10;
store D into '/user/zkpk/tmall/r2';
```

job.properties 文件内容如下:

```
nameNode=hdfs://master:9000
jobTracker=master:18040
queueName=default
examplesRoot=examples
oozieRoot=oozie
```

```
oozie.use.system.libpath=true
oozie.libpath=share/lib/lib_20170101021854/pig
oozie.wf.application.path=${nameNode}/user/zkpk/tmall/q2
```

文件上传完毕后,使用 oozie 命令执行工作流。

```
[zkpk@master oozie-4.2.0]$bin/oozie job -oozie
http://master:11000/oozie -config
/home/zkpk/tmall/q2/job.properties   -run
```

等待运行完成之后,可以在 MySQL 中查看结果。

```
mysql> select * from r2 order by count+0 desc;
+-------------+-------+
| pid         | count |
+-------------+-------+
| 17359010576 | 41561 |
| 17525034357 | 33424 |
| 35570005627 | 13974 |
| 38330727924 | 4720  |
| 24965336234 | 4709  |
| 37961093174 | 4655  |
| 41379268069 | 4521  |
| 36826954133 | 4049  |
| 37960797318 | 3994  |
| 38644070489 | 3571  |
+-------------+-------+
10 rows in set (0.03 sec)
```

2. 网站销售类分析

(1) 一段时间内的总销售额

使用 Oozie 工作流,workflow.xml 文件内容如下:

```xml
<workflow-app name="weibo_wf" xmlns="uri:oozie:workflow:0.5">
    <start to="query" />
    <action name="query">
        <pig>
            <job-tracker>${jobTracker}</job-tracker>
            <name-node>${nameNode}</name-node>
            <configuration>
                <property>
                    <name>oozie.launcher.mapred.child.java.opts</name>
                    <value>-Xmx2048m</value>
                </property>
                <property>
                    <name>pig.spill.extragc.size.threshold</name>
                    <value>100000000</value>
                </property>
                <property>
                    <name>mapred.child.java.opts</name>
                    <value>-Xmx2048m</value>
                </property>
                <property>
                    <name>mapred.user.jobconf.limit</name>
                    <value>100000000</value>
                </property>
```

```xml
                    </configuration>
                    <script>q3.pig</script>
                    <param>date1=2014-12-11</param>
                    <param>date2=2014-12-13</param>
                </pig>
                <ok to="export" />
                <error to="fail" />
            </action>
            <action name="export">
                <sqoop xmlns="uri:oozie:sqoop-action:0.2">
                    <job-tracker>${jobTracker}</job-tracker>
                    <name-node>${nameNode}</name-node>
                    <configuration>
                        <property>
                            <name>mapred.job.queue.name</name>
                            <value>${queueName}</value>
                        </property>
                    </configuration>
                    <command>export --connect "jdbc:mysql://master:3306/tmall" --
                    username hadoop --password hadoop --table r3 --input-fields-
                    terminated-by '\t' --export-dir /user/zkpk/tmall/r3</command>
                </sqoop>
                <ok to="end" />
                <error to="fail" />
            </action>
            <kill name="fail">
                <message>error
                    message[${wf:errorMessage(wf:lastErrorNode())}]</message>
            </kill>
            <end name="end" />
</workflow-app>
```

q3.pig 文件内容如下：

```
data=load '/user/zkpk/tmall/data/tmall-201412.csv' as(uid,time:chararray,pname,
price:double,number:int,pid);
A=filter data by time >'$date1' and time<'$date2';
B=group A all;
C=foreach B generate '$date1','$date2',ROUND(SUM(A.price));
store C into '/user/zkpk/tmall/r3';
```

job.properties 文件内容如下：

```
nameNode=hdfs://master:9000
jobTracker=master:18040
queueName=default
examplesRoot=examples
oozieRoot=oozie

oozie.use.system.libpath=true
oozie.libpath=share/lib/lib_20170101021854/pig
oozie.wf.application.path=${nameNode}/user/zkpk/tmall/q3
```

文件上传完毕后,使用 oozie 命令执行工作流。

```
[zkpk@master oozie-4.2.0]$bin/oozie job -oozie
http://master:11000/oozie -config
/home/zkpk/tmall/q3/job.properties  -run
```

等待运行完成之后,可以在 MySQL 中查看结果。

```
mysql> select * from r3;
+------------+------------+---------+
| date1      | date2      | sum     |
+------------+------------+---------+
| 2014-12-11 | 2014-12-13 | 1799732 |
+------------+------------+---------+
1 row in set (0.00 sec)
```

(2) 一段时间内的总订单量

使用 Oozie 工作流,workflow.xml 文件内容如下:

```xml
<workflow-app name="weibo_wf" xmlns="uri:oozie:workflow:0.5">
    <start to="query" />
    <action name="query">
        <pig>
            <job-tracker>${jobTracker}</job-tracker>
            <name-node>${nameNode}</name-node>
            <configuration>
                <property>
                    <name>oozie.launcher.mapred.child.java.opts</name>
                    <value>-Xmx2048m</value>
                </property>
                <property>
                    <name>pig.spill.extragc.size.threshold</name>
                    <value>100000000</value>
                </property>
                <property>
                    <name>mapred.child.java.opts</name>
                    <value>-Xmx2048m</value>
                </property>
                <property>
                    <name>mapred.user.jobconf.limit</name>
                    <value>100000000</value>
                </property>
            </configuration>
            <script>q4.pig</script>
            <param>date1=2014-12-11</param>
            <param>date2=2014-12-13</param>
        </pig>
        <ok to="export" />
        <error to="fail" />
    </action>
    <action name="export">
```

```xml
<sqoop xmlns="uri:oozie:sqoop-action:0.2">
    <job-tracker>${jobTracker}</job-tracker>
    <name-node>${nameNode}</name-node>
    <configuration>
        <property>
            <name>mapred.job.queue.name</name>
            <value>${queueName}</value>
        </property>
    </configuration>
    <command>export --connect "jdbc:mysql://master:3306/tmall" --username hadoop --password hadoop --table r4 --input-fields-terminated-by '\t' --export-dir /user/zkpk/tmall/r4</command>
</sqoop>
<ok to="end"/>
<error to="fail"/>
    </action>
    <kill name="fail">
        <message>error
            message[${wf:errorMessage(wf:lastErrorNode())}]</message>
    </kill>
    <end name="end"/>
</workflow-app>
```

q4. pig 文件内容如下：

```
data=load '/user/zkpk/tmall/data/tmall-201412.csv' as(uid,time:chararray,pname,price:double,number:int,pid);
A=filter data by time>'$date1' and time<'$date2';
B=group A all;
C=foreach B generate '$date1','$date2',SUM(A.number);
store C into '/user/zkpk/tmall/r4';
```

job.properties 文件内容如下：

```
nameNode=hdfs://master:9000
jobTracker=master:18040
queueName=default
examplesRoot=examples
oozieRoot=oozie

oozie.use.system.libpath=true
oozie.libpath=share/lib/lib_20170101021854/pig
oozie.wf.application.path=${nameNode}/user/zkpk/tmall/q4
```

文件上传完毕后，使用 oozie 命令执行工作流。

```
[zkpk@master oozie-4.2.0]$bin/oozie job -oozie
http://master:11000/oozie -config
/home/zkpk/tmall/q4/job.properties  -run
```

等待运行完成之后，可以在 MySQL 中查看结果。

```
mysql> select * from r4;
+------------+------------+-------+
| date1      | date2      | sum   |
+------------+------------+-------+
| 2014-12-11 | 2014-12-13 | 41175 |
+------------+------------+-------+
1 row in set (0.00 sec)
```

本章小结

本章通过两个案例介绍了整合使用各类离线处理组件完成数据分析的流程，读者可以对数据分析的需求分析有所了解，同时能够使用相关组件完成这些分析需求。

参考文献

[1] 卡普廖洛,万普勒,卢森格林. Hive 编程指南[M]. 曹坤,译. 北京:人民邮电出版社,2013.
[2] Alan Gates. Pig 编程指南[M]. 曹坤,译. 北京:人民邮电出版社,2013.
[3] Boris Lublinsky,Kevin T. Smith,Alexey Yakubovich. Hadoop 高级编程[M]. 北京:清华大学出版社, 2014.
[4] 黄宜华. 深入理解大数据[M]. 北京:机械工业出版社,2014.
[5] Alex Holmes. Hadoop 硬实战[M]. 北京:电子工业出版社,2015.
[6] 陆嘉恒. Hadoop 实战[M]. 2 版. 北京:机械工业出版社,2012.